Interdisciplinarity and Climate Change

Interdisciplinarity and Climate Change is a major new book addressing one of the [illegible] questions of our time. Its unique standpoint is based on the [illegible] coherent interdisciplinarity is necessary to deal w[illegible] multitude of linked phenomena which both constitu[illegible]

In the opening chapter, Roy Bhaskar makes use of [illegible] to articulate a comprehensive framework for multi[illegible], trans-disciplinarity and cross-disciplinary understanding, one which duly takes account of ontological as well as epistemological considerations. Many of the subsequent chapters seek to show how this general approach can be used to make intellectual sense of the complex phenomena in and around the issue of climate change, including our response to it.

Among the issues discussed, in a number of graphic and compelling studies, by a range of distinguished contributors, both activists and scholars, are:

- The dangers of reducing all environmental, energy and climate gas issues to questions of carbon dioxide emissions
- The problems of integrating natural and social scientific work and the perils of mono-disciplinary tunnel vision
- The consequences of the neglect of issues of consumption in climate policy
- The desirability of a care-based ethics and of the integration of cultural considerations into climate policy
- The problem of relating theoretical knowledge to practical action in contemporary democratic societies

Interdisciplinarity and Climate Change is essential reading for all serious students of the fight against climate change, the interactions between public bodies, and critical realism.

Roy Bhaskar is the originator of the philosophy of critical realism and the author of many acclaimed and influential works, including *A Realist Theory of Science*, *The Possibility of Naturalism*, *Scientific Realism and Human Emancipation*, *Reclaiming Reality*, *Philosophy and the Idea of Freedom*, *Dialectic: The Pulse of Freedom*, *Plato Etc.*, *Reflections on meta-Reality* and *From Science to Emancipation*. He is an editor of *Critical Realism: Essential Readings* and was the founding chair of the Centre for Critical Realism. Currently he is a World Scholar at the University of London Institute of Education.

Cheryl Frank was educated at the University of Illinois, earning master's degrees in political science and journalism. Her current interests include relating the philosophy of critical realism and meta-Reality to trends in British cultural studies and critical discourse analysis, especially in the fields of environmental education and peace studies.

Karl Georg Høyer is Professor and Research Director at Oslo University College. He holds a master's degree in technology and a PhD in social sciences with a dissertation on "Sustainable Mobility". Most of Høyer's research is related to sustainable development, with a main focus on transport and energy.

Petter Næss is Professor in Urban Planning at Aalborg University, Denmark, with a part-time position at Oslo University College, Norway. His main research interests are land use and travel; impacts and driving forces of urban development; philosophy of science. His most recent book is *Urban Structure Matters* (Routledge, 2006).

Jenneth Parker has linked interests in ethics, science, social movements and knowledge and has worked with Education for Sustainability at London South Bank University, WWF-UK, Science Shops Wales and UNESCO. She is a Research Fellow at the GSOE, University of Bristol, working on interdisciplinarity and sustainability/climate change.

Interdisciplinarity and Climate Change

[illegible] book addressing one of the most challenging [illegible] based on the recognition that effective and [illegible] with the issue of climate change, and the [illegible] and connect them.

[illegible] the extensive resources of critical realism [illegible] interdisciplinarity, transdisciplinarity [illegible] which duly takes [illegible]

[illegible]

Interdisciplinarity and Climate Change

Transforming knowledge and practice for our global future

**Edited by Roy Bhaskar,
Cheryl Frank, Karl Georg Høyer,
Petter Næss and Jenneth Parker**

LONDON AND NEW YORK

First published 2010
by Routledge
2 Park Square, Milton Park, Abingdon, Oxon, OX14 4RN

Simultaneously published in the USA and Canada
by Routledge
270 Madison Avenue, New York, NY 10016

Routledge is an imprint of the Taylor & Francis Group, an informa business

Typeset in Goudy by
Swales & Willis Ltd, Exeter, Devon
Printed and bound in Great Britain by
Antony Rowe, Chippenham, Wiltshire

British Library Cataloguing in Publication Data
A catalogue record for this book is available from the British Library

Library of Congress Cataloging-in-Publication Data
A catalog record has been requested for this book

ISBN10: 0–415–57387–4 (hbk)
ISBN10: 0–415–57388–2 (pbk)
ISBN10: 0–203–85531–0 (ebk)

ISBN13: 978–0–415–57387–0 (hbk)
ISBN13: 978–0–415–57388–7 (pbk)
ISBN13: 978–0–203–85531–7 (ebk)

Contents

Introduction

This book represents a dynamic engagement between interdisciplinary approaches to one of the major issues of our time and the philosophy of critical realism. Contributions in this book are all inspired by a commitment to interdisciplinary approaches and analysis, and many of the contributions employ a critical realist framing of the issues at stake. The extensive resources of critical realism are outlined in relation to climate and its interdisciplinary nature in this book in various ways. Strong arguments are presented to show that critical realist approaches, or something very close to them, will be an indispensable part of an adequate intellectual response to climate change and the multitude of linked phenomena with which we have to deal in the twenty-first century. The radical inadequacy of piecemeal approaches to our joined-up world is presented on every page of this book – however, positive indications of more integrative ways forward are also presented. Crucially, critical realism demonstrates that it is not enough to have a metaphysical disposition to take a joined-up view; intellectual tools are required to help us handle this task which is hugely challenging and should not be underestimated. The discussion and elaboration of some of the tools that we need are the contribution of this book.

Climate change is recognised by many as a crisis that is calling into question our whole approach to development – this book argues that it must also be seen as calling into question the ways in which we develop and use knowledge. Even those who see climate change as an urgent issue, for the most part, lack a framework for coherently integrating the findings of distinct sciences, on the one hand, and for integrating those findings with political discourse and action, on the other. This volume addresses a wide sweep of these issues of integration, ranging from integration across (relatively) adjacent sciences; between physical sciences and social sciences; to case studies focusing on key areas of climate-related policy, such as energy technology debates; ways to conceptualise and measure relationships between social activities and climate outcomes in pursuit of reductions in greenhouse gases; and thematic studies of strongly climate-related issues such as food crises. In addition, this volume contains a number of detailed critiques of the undermining effects of lack of integration in some crucial fields of knowledge such as planning, economics and the policy/civil society interface in relation to climate change.

True to the dialectical impulse, the ways in which studying and responding to major systemic phenomena across a range of domains of reality also create new challenges for philosophy, strategy, policy and action, are considered. Diversity and interdisciplinarity has always been a strength of critical realism, with conferences, colloquia and meetings ranging from the annual conferences of the International Association of Critical Realism to the meetings of specific research networks, representing a wide range of diverse disciplinary areas in addition to more generalized philosophical developments and critique. In this spirit, identifying areas for future research for the critical realist programme is also an important intellectual outcome of this volume. There is also an important link between theory and practice in that those who are at the forefront of developing interdisciplinary research and practice help to identify problems and issues that constitute a challenge for theory, but also help to illustrate theoretical problems in illuminating ways. In addition, critical realist engagement with other areas of thought that have contributed to thinking in this area, such as systems theory, can be a rich source of future dialogue and possible development.

The stress on active interdisciplinary working of research and policy councils is a relatively new emphasis and the evidence is that academic communities are struggling to respond. The extent to which a joined-up world needs joined-up knowledge and practice is being urgently reviewed throughout health, child welfare and education, in addition to the vital recognition of the relative fragility of the linked life support systems of the planet in the face of climate change and the demands of a rapidly increasing global population. In civil society these moves are also evident. For example, as NGOs and civil organisations perceive the need to link up environment, human development and care issues more fully, they also need the tools and thinking to enable them to do so effectively. Those who are trying to engage wider civil society are also faced with a key problem – how can we integrate information from different disciplinary sources into pictures that make sense to people sufficiently to inform their decisions? Critical realism – as a philosophical framework encompassing an ontology that ranges from the metatheory of so-called hard science through biology and evolutionary theory, to social sciences, to a critical engagement with the 'cultural turn' and the importance of discourse to human action and identity and action – is a good candidate to help to 'broker' interdisciplinary approaches.

The book's unique standpoint stems from the fact that critical realism, or something very close to it, is required to show both why interdisciplinarity is necessary, and how it, together with interprofessional cooperation generally, is possible in practice. The first chapter, by Roy Bhaskar, succinctly restates and rearticulates the theory of multidisciplinarity, interdisciplinarity, transdisciplinarity and cross-disciplinary understanding (and inter-professional cooperation) developed by Roy Bhaskar and Berth Danermark in their seminal 2006 article.[1] Many of the subsequent chapters in *Interdisciplinarity and Climate Change* (IDCC) explore the ways in which the conceptual framework developed by Bhaskar and Danermark, and that of critical realism generally (including not only basic or original critical realism, but also dialectical critical realism and the philosophy of meta-Reality) can cast illuminating light on contemporary problems of understanding and dealing with climate change.

Chapter 1, 'Contexts of interdisciplinarity', by Roy Bhaskar, argues that only a comprehensive and articulated interdisciplinary approach can do justice to pressing questions of climate change; and that the philosophical approach of critical realism, or something equivalent to it, is required to intellectually sustain and practically develop such an interdisciplinarity. That is to say, critical realism is uniquely capable of situating the weaknesses of actualist, reductionist, monodisciplinary accounts of science, and the necessity for interdisciplinary work in dealing with complex concrete phenomena such as climate change.

In the first part of the chapter, after elucidating the basis of disciplinarity in science, Bhaskar rehearses the progressive argument for multidisciplinarity, interdisciplinarity, transdisciplinarity and cross-disciplinary understanding. The resulting concept of a laminated system pinpoints the meshing of explanatory mechanisms at several different levels of reality and possible orders of scale. The chapter then goes on to consider the articulation of laminated systems, making use of the expanded conceptual frameworks of dialectical critical realism and the philosophy of meta-Reality. Turning to the social domain, the chapter argues for the necessity of a conception of four-planar social being, at potentially up to seven orders of scale, and for a view of social life as concept dependent but not concept exhausted, so paving the way for critical discourse analysis. Having developed the concepts necessary for the reconstruction of contemporary discourse on climate change, Bhaskar turns to the forms of its critique, including immanent, ommisive and explanatory critique and rearticulates a standpoint of concrete utopianism, arguing that a key role for intellectuals consists in the envisaging of alternative possible futures for humanity.

Chapter 2, by Sarah Cornell and Jenneth Parker, applies the argument and conceptual framework developed in Chapter 1 for complex concrete phenomena in general to the specific case of climate change, illustrating Bhaskar's argument. Together, Chapters 1 and 2 set the agenda for the specific studies in the remainder of *Interdisciplinarity and Climate Change*.

In a discussion of great moment, Karl Georg Høyer argues in Chapter 3 that the current focus on efforts to mitigate climate change is dominated by a particular form of reductionism – this is carbon, and even more especially, carbon dioxide (CO_2) reductionism, a reductionism which encompasses three distinct levels, successively embracing the reduction of all climate gases, then all energy issues, thence all environmental issues, to CO_2. Høyer then proposes seven theses to move away from such reductionism as a basis for more credible mitigation efforts. These include the need to reduce energy consumption, economic volumes and consumption volumes (on the basis of a systematic differentiation between issues of volume, distribution and allocation). The chapter concludes in a powerful concrete utopian call for substantive visions of a 'post-carbon society'.

In a meticulously argued and insightful chapter on 'The dangerous climate of disciplinary tunnel vision', Petter Næss shows that, while theories and their applications in energy and climate studies need to be strongly based on interdisciplinary integration, such holistic approaches are rare in both academic and political discourse. The chapter then traces some possible reasons why monodisciplinary reductionist approaches are so prevalent in spite of their serious shortcomings, which Næss systematically details. He pays particular attention, on the one hand, to the role of non-critical realist (e.g. positivist or strong social constructivist) metatheoretical positions in explicitly excluding both certain types of knowledge and the methods necessary for multidimensional analysis; and, on the other hand, the politico-economic interests potentially threatened by consideration of the relationships between neo-liberal policies and climate change crisis. Næss concludes with an alternative storyline incorporating insights from interdisciplinary research omitted by currently dominant mainstream storylines.

Chapter 5 by Carlo Aall and John Hille provides an important corrective for contemporary discussions of climate policy, in which consumption is indeed, as their title suggests, a 'missing dimension'. Dividing greenhouse gas emissions into two groups, those caused by consumption and those caused by production, Aall and Hille show that emissions in developed countries are increasingly related to the consumption and not to the production of goods and services. They then discuss the need to develop a more comprehensive and consumption-related climate policy approach to the mitigation of greenhouse gas emissions.

Chapter 6 by Cheryl Frank looks at the cultural articulations and social imaginaries around global warming. She argues that critical realism provides a comprehensive and inclusive framework and set of tools for addressing the very complex phenomena of global warming and climate change generally, including its socio-cultural landscapes. She shows how the theory of articulation developed by Stuart Hall and members of the British cultural studies school, taken together with the insights of Antonio Gramsci, Raymond Williams and other cultural theorists, can substantially augment our understanding of the issues necessary to comprehend and decisively tackle climate change on the various levels and scales of planetary life. Making a fruitful connection between the theory of articulation and contemporary critical discourse analysis, she then develops some of the crucial elements for a critical realist cultural theory of social semiosis or meaning, arguing that something like 'articulated laminated systems' must be identified as the indispensable units for action on climate change. Finally, Frank turns to the way in which local knowledge and wisdom can be recuperated and integrated into an emerging 'zyxa formation',[2] that is, a social imaginary based on optimism of the will and realism – not pessimism – of the intellect, informed by concrete utopianism.

In the next chapter, Sarah Cornell gives an absorbing historical account of the formation of contemporary climate science, from its parent subjects oceanography and meteorology, through the mid twentieth century codification of understanding in models and system science. While contemporary earth system models now produce awe-inspiring results, the uneasy co-existence of high certainty and deep uncertainty in our understanding of climate has definite political effects. Cornell argues that physical science has reached its explanatory limits in the climate context and now needs to be integrated with the human sciences, a project which it has been reluctant to undertake and for which a critical realist perspective is essential. Current divisions of climate issues (e.g. by the IPCC) into separate study areas continue to reflect and reinforce traditional disciplinary (e.g. science/arts) cultural divides in climate research.

Chapter 8 is on the terrain of economics. Writing from the perspectives of ecological economics, Robert Costanza argues that the mainstream model of development is based on a number of antiquated assumptions about the way the world works. In the contemporary context, characterized by climate crisis, we have to reconceptualize the nature of the economy. We need to remember that the goal of the economy is to improve human well-being and quality of life, that material consumption and GDP are merely means to that end, not ends in themselves. We have to better understand what really does contribute to sustainable human well-being, and recognize the substantial contributions of natural and social capital. We need to be able to distinguish between real poverty in terms of low quality of life, and merely low monetary income. Ultimately, we have to create a new vision of what the economy is and what it is for, and a new model of development, which acknowledges this new context and vision. This will require the full engagement of economics with other disciplines.

Chapter 9 of *Interdisciplinarity and Climate Change*, by Kjetil Rommetveit, Silvio Funtowicz and Roger Strand, looks at how the relationship between knowledge and action is conceived in modern 'knowledge-based' societies. The authors analyze a situation in which, while it is clear from countless reports (IPCC, Stern, etc.) that it is irresponsible to question the seriousness of the situation, governments of all complexions hesitate to implement climate policies that respond to the dramatic threats indicated by these reports. The authors argue that the climate issue is becoming deeply emblematic for global problems in general, in which 'stakes are high, decisions are urgent, facts are contested and uncertainty cannot be eliminated', and go on to consider how we are to arrive at effective climate policies in democratic societies, in which critical and sceptical voices cannot be silenced and doubt can never be entirely eliminated *ex ante*.

In Chapter 10, Karl Georg Høyer puts the conceptual resources of critical realism and the theory of interdisciplinarity proposed in Chapter 1 to work to develop a concept of technological idealism in the analysis of the recent nuclear power debate in Norway. The context of this was the need to develop carbon-free energy production, both nationally and globally. In the debate a complete change in nuclear power technology came to be envisaged. This was termed ADS (standing for accelerator-driven systems) and based on thorium, which Norway possessed in large quantities, rather than uranium. The prospectus presented painted a picture of Norway in a world-leading position developing this technology, which was claimed to have huge ecological and economic benefits. Høyer systematically exposes the methodological and substantive flaws of the arguments put forward by the protagonists of thorium in the nuclear power debate in Norway.

In the next chapter, Hugh and Maria Inês Lacey turn their attention to the relationship between the contemporary food crisis and issues of global warming, and more generally, climate change from a critical realist and interdisciplinary perspective. They show that the mechanisms explaining the contemporary food crisis are rooted in a capital-intensive, industrial and corporate form of agricultural production, systematically integrated into the global market and its institutions; heavily dependent on petro-chemical inputs and techno-scientific innovations; and implemented by way of soil-depleting planting monocultures. The explanatory critique of such agri-business points to the necessity for an alternative conception, based on local food sovereignty. The authors detail the kind of interdisciplinary investigations necessary for the careful design of such an agroecosystem, rich in biodiversity and yielding a portfolio of products. Such a system eliminates much of the need for chemical fertilizers, herbicides and pesticides and results in the production of food under conditions in which sustainability and social health are strengthened and rated more highly than profit and economic growth.

In Chapter 12, Jenneth Parker outlines ways in which the resources of a dialectically conceived critical realist interdisciplinarity can combine with some aspects of communitarian, feminist and ecofeminist ethics and considers how interdisciplinary understandings of the human condition and of our concrete embodied singularity help us to overcome the dilemma of universal contextual ethics. She then employs a dialectical critical realist framework to argue the case for a new humanism based on care, including, specifically, care of the environment. Parker argues that a concern with care has been a chief driver of interdisciplinarity and its discourses; and that care, on all four planes of social being, while compatible with an overreaching philosophical non-anthropocentricity, can form the basis for a new immanent humanist ethic that is capable of sustaining a continuing commitment to human emancipation and self-realization, rather than just mere survival. In the course of this argument, she makes the important point that our responses to climate change may be in terms of one or more of the three modalities of mitigation, adaptation and regeneration or restoration.

In the epilogue, Karl Georg Høyer elaborates on the paradoxes and dilemmas of conference tourism. Writing in the laconic style of Norwegian eco-philosophy, the author argues that conference tourism is part of the globalization of academia, producing little or nothing of lasting value, but generating in its wake serious and deleterious ecological effects. Such conference tourism is only a part of global tourism. In no other field, he argues, are there larger differences in ecological loads between the highly mobile global elite and the vast immobile majority of the world population.

Roy Bhaskar and Jenneth Parker
(on behalf of the editors)
September 2009

Notes

1 Roy Bhaskar and Berth Danermark, 'Metatheory, interdisciplinarity and disability research: a critical realist perspective', *Scandinavian Journal of Disability Research*, Vol. 8, No. 4 (2006).
2 See Mervyn Hartwig, *Dictionary of Critical Realism*, Routledge 2007, p. 503.

1 Contexts of interdisciplinarity

Interdisciplinarity and climate change

Roy Bhaskar

This chapter is concerned with exemplifying the triangular relationship of critical realism, interdisciplinarity and complex (open-systemic) phenomena such as climate change. Whereas this chapter will consider these relationships in an abstract and general form, Chapter 2 will consider the application of the argument of this chapter to climate change in more specific detail. To some (varying) extent, also the other chapters in this book will exemplify various aspects of the argument developed here. This chapter is necessarily somewhat summary and abstract, but for a fuller development of the general argument, see Part 1 of my forthcoming book with Berth Danermark, *Being, Interdisciplinarity and Well-Being*.[1]

The core argument of original critical realism

What has been called 'original', 'basic' or 'first wave' critical realism was constructed on a double argument from experimental and applied activity in natural sciences such as physics and chemistry. This double argument was, on the one hand, for the revindication of ontology, or the philosophical study of being, as distinct from and irreducible to epistemology, or the philosophical study of knowledge; and, on the other hand, for a new radically non-Humean ontology allowing for structure, difference and change in the world, as distinct from the flat uniform ontology implicit in the Humean theory of causal laws as constant conjunctions of atomistic events or as invariant empirical regularities – a theory which underpins the doctrines of almost all orthodox philosophy of science.[2] This argument situated in the first place

- the necessity to disambiguate ontology and epistemology, based on a critique of what I called the epistemic fallacy (or the analysis or reduction of being to knowledge of being);
- the necessity, accordingly, to think science in terms of two dimensions, the intransitive dimension of the being of objects of scientific investigation and the transitive dimension of socially produced knowledge of them; and
- the compatibility of ontological realism, epistemological relativism and judgmental rationality, the 'holy trinity' of critical realism.

At the same time, it also situated the necessity to think of reality in terms of at least three domains, the domains of the real, the actual and the empirical, with the real encompassing the actual and the empirical, but also including non-actualized possibilities or powers and liabilities, either as unmanifest or as exercised though not actualized in a particular sequence of events, and where the actual includes the empirical but also things and events which exist or occur unperceived or more generally unexperienced by human beings. This latter aspect of the argument generated a critique of the actualism and reductionism prevalent in contemporary philosophies of science (and social science).

The critical argument, or means of establishing these cardinal propositions, in basic or original critical realism, depended on the observation that, outside experimentally established and a few naturally occurring 'closed' contexts, invariant empirical regularities do not occur. The need in general to artificially generate them means that they cannot be identified with the causal laws and other objects of scientific knowledge that they ground, but must be seen as our mode of empirical access to them; and that the causal laws, etc. must be analysed as objects which exist and act independently of our access to them, including transfactually (i.e. outside the context of their establishment). They must therefore be conceived as the operation of structures and mechanisms which exist and act independently of our human (experimental) access to them.

This argument, together with complementary considerations relating to our applied activity, establishes the foundational double result of original critical realism, involving the critiques of the epistemic fallacy and of actualism in ontology. The limited but real basis of the epistemic fallacy lies in what I have called the natural attitude, i.e. the fact that we do not normally disambiguate ontological and epistemological questions in our ordinary discourse about the known world; and the limited but real basis of actualism (and hence reductionism) in ontology lies in the empiricist misconstruction of the experimental successes of the natural sciences. In other words, the basis of orthodox accounts of science lies in two fundamental category mistakes, which are isolated by basic or original critical realism.

However, it is important to note that ontology is always in principle distinct from epistemology, even where our knowledge of the known world is unquestioned; and that structures, mechanisms, processes, fields and the other intransitive objects of scientific knowledge are always distinct from, and irreducible to, the patterns of events they generate, even in experimentally closed laboratory situations.

In principle then, we must always distinguish between, for example (a report, statement or claim about):

A1 'the distance between Rio de Janeiro and Florianopolis'; and
(a report, statement or claim about)

A2 'our knowledge about the distance between Rio de Janeiro and Florianpolis'; and between

B1 'the relationship between two measured variables (or experienced events)';
B2 'the pattern yielded by two events (or types of events)'; and
B3 'what it is (i.e. the structure or mechanism, etc.) that when stimulated, released or triggered by the first event or type of event generates, or tends to generate, the second event or type of event'.

It is an implication of this argument that, outside experimentally and a few naturally occurring closed contexts, the world is constituted by open systems. The resulting account of science emphasizes in particular three aspects or senses in which the world, and science accordingly, is stratified. There is

- a distinction between structures and events, or the domains of the real and the actual;
- the reiterated application of this distinction in a conception of the world as consisting of multiple layers of such strata, i.e. it reveals a multi-tiered stratification (material objects such as tables and chairs are constituted by molecules, which are in turn constituted by atoms, which are, in turn, constituted by electrons, which are, in turn, constituted by more basic phenomena or fields); and
- the conception that among such strata are levels characterized by the striking phenomenon of emergence.

Here an emergent level is understood:

- as unilaterally dependent on a more basic level;
- as taxonomically irreducible to it; and, most importantly,
- as causally irreducible.

A characteristic pattern of discovery and theoretical explanation follows from this ontology and account of science as consisting essentially in the movement from events to the structures that generate them. This defines a characteristic logic of scientific discovery, involving what I have called the DREIC schema, where D stands for the description of some pattern of events or phenomena; R for the retroduction of possible explanatory mechanisms or structures, involving a disjunctive plurality of alternatives; E for the elimination of these competing alternatives; I for the identification of the causally efficacious generative mechanism or structure; and C for the iterative correction of earlier findings in the light of this identification.

The implications of open systems

If in Section 1 of this chapter, I have been in effect elaborating (or rather rehearsing the elaboration of) the case for *disciplinarity* in science, viz in the specialized creative and transformative work (in the transitive dimension) necessary for the identification of the previously unknown deep structures and *causal* mechanisms of the world (in the intransitive dimension), I now turn to development and examination of the case, likewise primarily ontological, for *interdisciplinarity*. The (philosophical) ontological nature of the case for interdisciplinarity developed here differentiates it from most of the literature in the field, which is overwhelmingly epistemologically (and sociologically) orientated.

Almost all the phenomena of the world occur in *open systems*. That is to say, unlike the closed systemic paradigm, they are generated not by one, but by a multiplicity of causal structures, mechanisms, processes or fields. A characteristic pattern for the analysis of explanation of such phenomena was developed in basic critical realism.[3] This involves 'the *RRREIC* schema', where the first R or R_1 stands for the *resolution* of the complex event or phenomenon into its components; the second R or R_2 for the *redescription* of these components in an (ideally, optimally) explanatory significant way; the third R or R_3 for the *retrodiction* of these component causes to antecedently existing events or states of affairs; E for the *elimination* of alternative competing explanatory antecedents; I for the *identification* of the causally efficacious or generative antecedents; and C for the iterative correction of earlier findings in the light of an (albeit temporarily) completed explanation or analysis.

Analysis of R_1

I will organize my approach to such phenomena around the first three moments of this analysis. R_1 signals the characteristic complexity of open-systemic phenomena, and registers the need to refer to a multiplicity of (successively) causes, mechanisms and theories in the explanation of the phenomenon. What is involved here is typically a conjunctive multiplicity of components, i.e. component a and b and c, rather than the disjunctive plurality that is involved in, say, the retroductive moment of theoretical science, when it is a case of either mechanism a or b or c. The *conjunctive multiplicity* specifies, one could say, the logical form of the *open* systemic phenomenon, and paves the way for introducing consideration of the ontological case for multidisciplinarity and interdisciplinarity.

The analysis of an open-systemic phenomenon establishes the characteristic multiplicity of causes, and *a fortiori* mechanisms and therefore, potentially, theories (of these mechanisms). From this characteristic *multi-mechanismicity* we cannot, however, infer the need for multidisciplinarity. For this, a further ontological feature besides *complexity* is required: this is *emergence*, more specifically the emergence of levels. We now have *multidisciplinarity*, ontologically grounded in the need to refer to a multiplicity of mechanisms at different, including emergent, levels of reality.

This stage of the argument is consistent with a purely additive pooling of the results of the knowledge of the distinct mechanisms. However, what is typically involved in the open systemic case (when emergence applies) is not only an emergence of levels, but an emergent outcome of the intermeshing of the different mechanisms. This requires genuinely synthetic interdisciplinary work, involving the epistemic integration of the knowledges of the different mechanisms. (I will consider the social implications of this in a moment.)

We now have emergent levels and emergent outcomes. So far, we have been assuming that the mechanisms involved in the explanation of the open-systemic phenomenon are unaffected by their new context. But a moment's reflection on phenomena such as the production of sounds or marks in human speech and writing shows that this will often not be the case; that the mechanisms involved may be radically altered by the new synthesis or combination. When the mechanisms themselves change, and are thus emergent, we can talk of *intradisciplinarity* rather than *interdisciplinarity*.

Until now, the pertinent considerations have been ontological. However, in interdisciplinary work what will be required are new concepts, theories and modes of understanding. This will necessitate epistemological transdisciplinarity, involving the exploitation of pre-existing cognitive resources drawn from a wide variety of antecently existing cognitive fields in models, analogies, etc. Such transdisciplinarity in creative interdisciplinary work has seemed to some writers to involve breaking with the very notion of a discipline, to the extent that there has been talk of *postdisciplinarity*. For reasons to be given later, I am cautious about this claim. However, it is evident that what will be required for successful interdisciplinary work at the epistemological level will be at the very least the capacity of members of the interdisciplinary team to communicate effectively with each other in *cross-disciplinary* understanding. And this, together with the need for elements of creative transdisciplinarity, will necessitate a form of education and continuing socialization of the interdisciplinary research worker, very different from that involved in orthodox monodisciplinarity (more on this later).

Ontologically, the most important result of our analysis thus far is the need to understand a form of determination in reality, in which several irreducibly distinct mechanisms at different and potentially emergent levels are combining to produce a novel result. The different levels necessary for the understanding of the result may be conceived as interacting or coalescing in what I have called a laminated system or totality.[4]

There can be no *a priori* account of what levels or the number of levels that may be involved in any particular explanation, or indeed sphere of research. However, as a heuristic device, Berth Danermark and I undertook an investigation of disability research,[5] in which we argued that, in general, in disability research it was necessary to refer to (i) physical, (ii) biological, and more specifically physiological, medical or clinical, (iii) psychological, (iv) psycho-social, (v) socio-economic, (vi) cultural and (vii) normative kinds of mechanisms in order to achieve satisfactory explanations. We used the concept of a *laminated system* to ontologically underpin a critique of the history of disability studies as involving successively three forms of reductionism: medical reductionism, socio-economic reductionism and cultural reductionism. Karl Georg Høyer and Petter Næss have applied this kind of analysis to ecology generally[6] and Gordon Brown to education.[7]

Redescription

What the second moment of analysis, that of redescription, signifies is the need for a decision about the appropriate level of description of the component cause in terms of abstractness or concretion. Is the causally relevant fact about some incident at breakfast a lost material object, a lost packet of tea-bags, a lost packet of Earl Grey tea-bags, or a lost packet of Fair Trade Earl Grey tea-bags? Is the most explanatory significant description of what happened in Germany under Nazi rule the fact that the country was depopulated or that millions of people died or that millions of people were killed or that millions of people were massacred?[8] Which of the myriad possible levels of description of some economic phenomenon is the explanatorily crucial one?[9]

This second moment of applied analysis, can (as in the case of the first) be further deepened. In particular, not only is it the case that individual things and events must be explained in terms of the intrication of a multiplicity of explanatory mechanisms, but also they must be conceived as concrete universals and singulars. Every particular phenomenon which instantiates in some way a universal law does so concretely; and every particular thing or event always also instantiates some or other (concrete) universal. As I have argued, the minimum analysis for any concrete universal or singular is a multiple quadruplicity.[10] That is to say every concrete phenomenon (thing or event) must be analysed not only as:

1. instantiating (transfactually) universal laws, but as
2. constituted by particular specific mediations which differentiate it from others of its kind (for example, a woman may be a nurse, trade unionist, mother of three, fan of the Rolling Stones, etc.). Moreover, each instance of such a differentiated universal will be characterized by
3. a specific geo-historical trajectory. This will further particularize it from others of its kind; and each such geo-historically specific and mediated instance of a universal will also be
4. irreducibly unique.

The logic of the *concrete universal* = singular takes us in to the dialectical critical realist deepening of the ontology of basic critical realism, in particular to incorporate its 3L, third level or holistic deepening. This will be further discussed below. Suffice it to mention here that each particular concrete thing may also be conceived as a developing (partial) totality, with at least some of its changes being internally or endogenously generated.

Of the remaining moments of analysis – R_3, E, I, C – I can comment only briefly here on R_3. This refers to the retrodiction of antecedent states of affairs. However, this implies that the law-like operation of the mechanisms are known, which in the open and especially social world will often prove not to be the case. In such circumstances the applied explanatory task of discovering antecedent states of affairs, involving retrodiction, will have to go hand in hand with the explanatory theoretical task of discovering the nature of the relatively enduring generative mechanisms at work,[11] involving retroduction.

Deepening the logic of complexity

We have seen that the deepening of the logic of complexity means that not only must complex open-systemic phenomena be analysed in terms of a conjunctive multiplicity, but also that the component parts, under any particular description, may themselves be conceived as complex in the sense of the concrete universal, and as such, constituted as quadruple multiplicities, which may moreover be bound together as a developing partial totality.

Moreover, the various resolved components of a complex phenomenon must in general be themselves analysed *holistically*, i.e. precisely as components of the whole of which they are component parts. Thus we have the phenomena of holistic causality, and the constitution of events (the components of the complex phenomenon) as a *nexus* and of structures as a *system*[12]– for example, as in the levels of a laminated system! In these cases the combination coheres as a whole in as much as:

- the form of the combination causally codetermines the elements; and
- the elements causally co-determine (mutually mediate or condition) each other, and so causally co-determine the form.

Such holistic causality depends on internal relationality. An element a may be said to be internally related to an element b if b is a necessary condition for the existence of a, whether this relation is reciprocal, symmetrical or not. In general, complexes will be composed of both internal and external relations, i.e. they are 'partial totalities'.

Now these component parts are not only parts of a complex, they themselves must in general be analysed as complexes, themselves composed of component parts. As such, these components are subject to an internal as well as a, so to speak, external holistic necessity, namely, as themselves complex wholes as well as part of a complex whole. Further, in particular, especially in so far as they are to be conceived (under any description) as things, rather than merely as events (or changes in things), they are concrete universals = singulars. So we have the triple logic of inner complexity as involving components which are:

(i) in what I have called holistic *intra-action*[13] (to differentiate it from the normal external-relational connotations of 'interaction');
(ii) themselves complexes containing component parts, which may be in turn in holistic intra-action; and
(iii) under any particular description, concrete universals.

This is the general ontological form of the concrete, as the conjuncture or compound or condensate, involving the coalescing of forces (more or less bound in to a unity of many determinations).

But, in addition to this triune *inner complexity*, any such concrete complex or component will also reveal an *outer complexity* in the form of

(iv) a context, which normally influences or shapes it, as distinct from generating or determining it.

The importance of context in social life cannot be exaggerated. In general, we cannot specify the operation of a mechanism in abstraction from its *context* – how the mechanism acts depends upon its context; so that we need to think of the (mechanism.context) couple as the effective generative dyad in social life, i.e. as that which produces outcomes or tendencies to outcomes in social being.[14]

Finally, there is the particularly strong form of

(v) co-complexity or joint determination, in a particular field or domain.

This occurs when two mechanisms from closely related spheres, such as, for example, politics and economics, become knitted or knotted together, or effectively inter-defined, so that a change in one is *a fortiori* a change in the other. Such a knot may be formed by, for example, the ideas of bourgeois society or the knowledge-based economy. (This characteristic binding of structures has sometimes been called 'cross-disciplinarity', but in this chapter I am giving that term a different sense.)

The nature of laminated systems

The concept of a laminated system has been derived above from reflection on the implications of R_1. But while R_1 establishes the pattern of explanation in terms of a conjunctive multiplicity, and hence a laminated system, it leaves the nature of the laminated system, i.e. the form of articulation of the conjunctive multiplicity, including the patterns of dependency and interaction, in principle open.

Substantive *a posteriori* analysis will, in general, be needed to determine this. Thus following our investigation into the case of disability research (and the substantive research efforts of one of us in this field), Berth Danermark and I were able to arrive at a real definition of disability studies as an articulated lamination 'in relation to the experience, and perception of the experience, of some impairment or functional loss, which itself or the effects of which, require to be socially or psychologically assessed, compensated (or accepted), transcended, mediated or otherwise reflected'.[15] Such a real definition achieved after an analysis of a field shows the way in which it evades unprincipled eclecticism, as the concept of a laminated system enables it to avoid reductionism.

Implications of critical naturalism

Original or basic critical realism is also, of course, developed to incorporate the understanding of specifically social and more generally human phenomena. This involves registration of a series of ontological, epistemological, relational and critical differences between social and natural phenomena and contexts of explanation. Following the method of immanent critique, an independent analysis in the field of philosophy of social science and social theory allows the resolution of characteristic dichotomies or dualisms, between structure and agency, society and individual, meaning and law, reason and cause, mind and body, fact and value, and theory and practice. The resolution of these dualisms cannot be rehearsed here, but the upshot is that

- the antinomy of structure and agency is resolved in the transformational model of social activity, in a conception on which social structures always pre-exist human agency, but are reproduced or transformed only in virtue of it and in the course of ongoing social activity;
- the antinomy between society and individual is resolved in a relational conception of the subject matter of the social sciences, namely as consisting not in behavior, either individual or collective, but paradigmatically in the enduring relations between individuals; and
- the antinomy between meaning and law or hermeneutics and positivism is resolved in a notion of social life as conceptually dependent but not exhausted by conceptuality, and of conceptuality as providing the necessary hermeneutic starting point for social investigation, but as in principle corrigible.

This understanding of social life is in turn predicated on:

- understanding of human agency as dependent upon intentional causality or the causality of reasons;
- synchronic emergent powers materialism; and
- recognition of the evaluative and critical implication of factual discourse.

The transformational model of social activity may be further developed to generate the notion of four-planar social being. This specifies that every social event occurs in at least four dimensions, that of material transactions with nature; that of social interactions between humans; that of social structure proper; and that of the stratification of the embodied personality. These four planes constitute, of course, a necessarily laminated system of their own in so far as reference to any one level or dimension will also necessarily involve reference to the others.

In a similar way each social level involved in any applied explanation can not only be situated in the context of four-planar social being, but also in that of a hierarchy of scale, that is of more macroscopic or overlying and less macroscopic or underlying mechanisms. Thus we can define distinct levels of agency and collectivity with which social explanation may be concerned, including the following seven levels:

(i) the sub-individual psychological level;
(ii) the individual or biographical level;
(iii) the micro-level studied, for example, by ethnomethodologists and others;

(iv) the meso-level at which we are concerned with the relations between functional roles such as capitalist and worker or MP and citizen;
(v) the macro-level orientated to the understanding of the functioning of whole societies or their regions, such as the Norwegian economy;
(vi) the mega-level of the analysis of whole traditions and civilizations; and
(vii) the planetary (or cosmological) level concerned with the planet (or cosmos) as a whole.

In this way we can see that the multiplicity and complexity deriving from level, context and scale may each result in the constitution of a laminated system.

The conceptual features of social life may in turn be developed so as to include – most fully in critical discourse analysis – an account of discourse as both constitutive of and conditioned (or causally affected) by, and in turn conditioning (or causally affecting), the extra-discursive aspects of social life as unfolded over four-planar, seven-scalar social being.

Interim summary of the argument

Current metatheories and methodologies of science encourage an actualist and reductionist, monodisciplinary approach to phenomena such as climate change. Conversely, such phenomena can only be understood in terms of the intrication of several distinct explanatory mechanisms, operating at radically different levels of reality, including four-planar social reality, and orders of scale. These range from the cosmological, through the physical, chemical, geological, biological, ecological (including the ecology of functioning ecosystems, living organisms in their environment and of climatic systems), psychological, social and normative. Focusing on individual entities in their environment allows us to define a clear hierarchy in which a higher order level has as its condition of possibility a more basic lower order level.

Within the human social field, we can further differentiate human ecological (at the level of human life support systems), social–material, social–institutional and social–cultural forms and aspects of human social practices. The socio-material level is concerned with the production, consumption, care and settlement of groups or collectivities of living human beings in their environments; the socio-institutional level is concerned with social, economic, political, military (etc.), familial, educational and linguistic forms and structures; and the socio-cultural level includes scientific, artistic, ethical, religious and metaphysical, elite and popular modes of expression, learning and interaction.

From a philosophical point of view, we have seen that the situation of a multiplicity of mechanisms operating at radically different levels of reality and orders of scale presupposes that the systems in which the mechanisms act are open and that some of these mechanisms operate at levels which are emergent from others. This necessitates, at the very least, a *multidisciplinary* approach. However, the fact that the outcomes may themselves be emergent means that the additive pooling of the knowledge of the different disciplines will not be sufficient, and that instead what will be required will be the synthetic *interdisciplinary* integration of the knowledges of the different disciplines. The approach adopted here may be characterized as ontologically a *developing integrative pluralism.*[16]

Epistemologically, for the successful pursuit of such interdisciplinary work, we need in addition both transdisciplinarity, involving the potential creative employment of models, analogies and insights from a variety of different fields and disciplines; and cross-disciplinarity, involving the potential to empathize with and understand and employ the concepts of disciplines and fields other than one's own. This has radical implications for both the curriculum and pedagogy of higher education, and arguably also for secondary and even primary education.

Between original critical realism and dialectical critical realism

So far, we have been largely developing the implications of complexity in domains characterized by emergence. Not only must most open-systemic phenomenon be analysed or resolved into their component or constituent parts, and their production seen as accordingly constituted by an effective laminated totality of generative mechanisms and causes; but we have seen that open-systemic phenomena, things or events are themselves, under any one description, constituted as a multiplicity in terms of four (potentially holistically related) dimensions, i.e. as concrete singulars = universals. If the former aspect means that we see open-systemic phenomena as essentially conjunctures, compounds or condensates, the latter aspect involves a further deepening of our understanding of the logic of the open-systemic phenomena, so that the particular component or constituent part is itself conceived as a multiple quadruplicity, defining the four dimensions of axes of the concrete universal. Moreover, this complexity is further accentuated in the social domain by the necessity to see social phenomena as occurring along four planes, and at potentially seven orders of scale. Alongside the notions of four-planar and seven-scalar social being, we must conceive social being as constituted in part by discourses, which are causally interrelated to the extra-discursive aspects of social reality, such as oppressive power relations. Furthermore, the mechanisms involved in the laminated totality cited or involved in the explanation of a concrete open-systemic phenomenon may be themselves be ordinated in terms of hierarchies of levels of being. Such hierarchies may be organized in terms of individuals in their environments; but additionally in the social world, in terms of hierarchies of practices, such as the material, institutional, cultural and normative, as well as in terms of their aspects (as in four-planar social being), scale or discursive characteristics.

In the laminated explanation of some open-systemic phenomenon, we may distinguish, in principle, natural, social and mixed determinations, with mixed determinations defining the practical order. Intermediate and concrete sciences lie between the abstract sciences and the reconstructed concepts of concrete objects. Concrete sciences study the ensemble of epistemically significant features of a given thing, whereas intermediate sciences study the confluence of two or more orders or types of determination in a given kind of thing or system. Note that, whereas natural laws fix boundary (and natural phenomena, initial) conditions for the social natural sciences, such as social biology, it is economic (political, etc.) mechanisms that set the boundary (and social phenomena, initial) conditions for the natural social sciences, such as technology.

A cognitive field such as disability studies or climatology may be regarded as constituting both an

- intermediate science, in so far as it studies the convergence of distinct mechanisms, at different orders of determination; and a
- regional science, in so far as it may also demarcate qualitatively new, emergent orders of determination.

Moreover, in so far as natural and social kinds of determinations are both involved, we are concerned with

- mixed determinations, as part of the practical order.

The fact that, in a laminated totality, natural and social determinants will often both be present means that interdisciplinary work must in general employ *mixed methods*. However, even within the social domain, the conception of social life as dependent upon, but not exhausted by, conceptuality means that in principle mixed methods must be employed here too. Thus the incidence of unemployment in a particular locality is a phenomenon which can be counted and measured, alongside a qualitative assessment of the reasons for it, based on, say, interviews with local businessmen, etc.

The critical realist conception of emergence gives rise to two characteristic models of superstructure (and accordingly hierarchy). On the first, the higher order level provides the boundary conditions for the lower order or more basic level (as, for example, economics provides the boundary conditions for the operation of the physical principles governing machines). On the second, the lower order or more basic level provides the conditions of possibility or framework for the emergent or higher order level, as, for example, ecology specifies the conditions of possibility of human material practices. Both models may be combined in creatively defining the hierarchy of levels in some laminated totality.

Clearly, the existence of laminated, and especially necessarily laminated, systems raises a complex problem for the articulation of the different levels in an explanation. In what way is the distinctive contribution of the different levels to be brought out and communicated in a coherent narrative? One way is to trace the causal series as it actually happens, the diachronic pattern of causality. Another way is to begin with the most basic or rooting or grounding (e.g. physical) level. These will not necessarily be the same, since patterns of diachronic causality will not necessarily coincide with orders of synchronic emergence and dependence, conceived as mapping out the existential order of dependency of causal mechanisms. A further response might be to deny the linearity involved in a sequential narrative exposition and to use tables, pictures or simultaneous equations, but what will often be heuristically most convenient is to follow the implied ordination of the dominant theory of the day. Refuting or qualifying this through its more glaring lacunae will allow a heuristically convenient organization of the phenomenon through the immanently critical preferred hypothesis.

In everyday discourse in the pragmatics of explanation one will always be abstracting from some causally relevant features, that is, features of the total situation which, had they been different, would have prevented or modified the effect in question. One is typically looking for what makes a difference in the particular case, against a background of assumed normality, and given the purposes of the explanatory enquiry. That is to say, 'when something is cited as a cause it is being viewed as that element, paradigmatically an agent, in the total situation then prevailing which, from the point of view of the cause-ascriber, "so tipped the balance of events as to produce the known outcome"'.[17]

However, whatever the pragmatics of the actual imputation or citation of causes, a critical realist principle of sufficient reason,[18] positing the ubiquity of explanations for differences, entails that there will always be an ontologically determinate sequence in diachronic explanation, that is to say there is always a real world order, however much we choose to abstract from some of it intricacies or parameters. Similarly, the precise structural weight and influence and exact role, rank and causal importance of particular mechanisms in any one explanatory enquiry must always, in principle, be regarded as ontologically determinate, although there may be no way of making this epistemologically determinable, and there may be no way of deciding upon the best explanatory focus irrespective of context.

The deepening of ontology and dialectical critical realism

So far, we have been considering concepts drawn from the subsequent development of the core argument of original critical realism (e.g. in its social extension) or from original critical realism generally (e.g. in holistic causality) in a relatively ad hoc way. However, following the immanent logic of the development of critical realism through dialectical critical realism and the philosophy of meta-reality enables us both to deploy a far greater range of concepts in elucidating the nature of the mechanisms at the various levels of our laminated systems, and also to augment our conceptual resources in thinking the relations between the mechanisms of the laminated systems.

The deepening of ontology enables us to expand the range of categories, e.g. to take in absence and negativity, including contradiction, internal relations, etc., which may be employed in our laminated explanations of open-systemic phenomena. It also allows us to expand our sense of our possible purposes in explanation, so as to enable us to frame an expanded range of questions to guide our focus in the articulation of the laminated system to hand. This may include (in the case of dialectical critical realism) its heuristic usefulness in suggesting fruitful ways of, or perspectives for, articulating laminated systems in terms of patterns of dependency and interdependency (3L), or from the standpoint of sources and types of change (2E) or from the perspective of the orientation of public policy (4D). In fact, in elaborating the holisitic (on p. 7) and social (on p. 9) aspects of concrete open-systemic phenomena, we have already made substantive inroads into the territory of dialectical critical realism, at 3L and 4D, respectively. So I will be relatively brief with these regions of dialectical critical realism here, focussing instead mainly on 2E.

The deepening of ontology entailed by dialectical critical realism involves its extension from the understanding of being of

- Being as such, and as structured (1M); to
- Being as process (2E),
- Being as a whole (3L), and
- Being as incorporating transformative practice (human agency)(4D).

The further extension of ontology entailed by the philosophy of meta-Reality incorporates the further understanding of

- Being as incorporating reflexivity (and spirituality) (5A),
- Being as re-enchanted (6R), and
- Being as non-dual (or as involving essential unity) (7 Z/A).

The system of dialectical critical realism

Corresponding to the four terms of dialectical critical realism – of non-identity or difference and structure (1M), absence and negativity (2E), open totality (3L) and transformative praxis (4D) – we have distinct emphases in explanation. I have already noted the emphasis at 1M, that of a conjunctive multiplicity or laminated totality constituted by several ontologically distinct but interacting mechanisms. The emphasis at 3L is on their relationships of dependency and interdependency, and of their characteristic patterns of interaction and intra-action. Thus here we have the characteristic framework model of levels of being employed by Jenneth Parker in her analysis of the conditions for sustainability,[19] developed from the model of superstructure as intra-structure, as formed within its conditions of possibility. Corresponding to 2E, we have distinctive emphases in the diachronic process of change, in the formation and dissipation of laminated totalities. Thus we can differentiate dialectical and entropic types of change, together with various forms of stasis or reproduction, or differentiate evolutionary from revolutionary processes of change. Corresponding to 4D, we have a special interest in the extent to which human transformative praxis can play a role in influencing and modifying the laminated totalities at work in the social sphere.

The reality of absence and the fallacy of ontological monovalence

Just as basic critical realism turned on the isolation of two cardinal category mistakes, namely the epistemic fallacy and ontological actualism, dialectical critical realism isolates a third, namely ontological monovalence or the generation of a purely positive account of reality. In contrast to this, dialectical critical realism argues that absence is constitutively necessary for being. A world without absence, without boundaries, punctuations, spaces, and gaps between, within and around its objects would be a world in which nothing could have determinate form or shape, in which nothing could move or change, and in which nothing could be differentiated or identified.

The fact that, in principle, reality is at least bivalent, i.e. characterized by absence as well as presence, can be seen by invoking R.M. Hare's triptych of the phrastic, neustic and tropic.[20] Thus in principle then we must always distinguish between, for example (a statement or claim about)

C1 'The presence or absence of rain in Rio de Janeiro' (involving an operation on the phrastic or ontic content of the proposition) and (a statement or claim about)
C2 'The affirmation or denial of the presence or absence of rain in Rio de Janeiro' (an operation on the neustic) and (an invitation or injunction to)
C3 'Imagine (pretend, hypothesize, entertain/suppose, act on) the presence or absence of rain in Rio de Janeiro'.

This last involves an operation on the tropic, and is very important in the concrete utopian movement of thought. The important point here, however, is that we have in C1–C3 instances of positivity or negation at three different levels, involving negation within reality, negation within factual discourse and negation within speculative or fictional discourse.

In other words, there is a difference between

C'_1 'being in, or say travelling to, Rio de Janeiro'; and
C'_2 'being in, or say travelling to, a discourse (or statement) about Rio de Janeiro'; and
C'_3 'being in, or say travelling to, a play (or story) about Rio de Janeiro'!

Absence is not only necessary for being, but change, properly understood, presupposes absence, i.e. the coming into being of new properties or entities and the passing away from being of previously existing ones. Absence yields not only the clue to the vexed problem of dialectic, which may be seen as depending on the rectification of absence (omissions, incompleteness) in a move to greater generality, inclusiveness and coherence, but is necessary for a full understanding of intentional action. For agency is the absenting of absence and this generates an axiology of freedom conceived as depending upon the absenting of constraints and unwanted and unneeded sources of determination. Absence is the key or root concept of this group of categories which includes most importantly the idea of contradiction, and the idea of contradiction as ontological and not just epistemological.

Internal relationality

A third level of categories revolves around the idea of internal relations between elements, and includes conceptions of holistic causality and the concrete universal, which we have already touched on.

The central idea at 3L is that of internal relations. It may be illustrated by the relations between successive statements or speech acts within a discourse, or acts or episodes within a frame of social life generally. Thus:

D1 a statement or discourse about Rio de Janeiro is, or may be, internally related to
D2 a question about Brazil or travel, but it will not typically be internally related to
D3 a game of chess in Springfield, Illinois, or the installation, say, of a jacuzzi in downtown Oslo, or
D4 the onset of the Crimean War, or
D5 the import of bananas into Sweden.

Transformative praxis

A fourth range of concepts pivots around human agency, or the idea of transformed transformative praxis, and includes the notions of the irreducibility of agency, intentionality, reflexivity and spontaneity in social life, which must be conceived in terms of the idea of four-planar social being, which again we have already encountered.

Dialectical critical realism enables a vastly expanded range of concepts which may be invoked in the laminated totalities employed in the explanation of a complex open-systemic phenomenon in emergent domains. In particular the levels may be conceived as constituted by absence and omissions and various contradictions, as internally as well as externally related, as concrete and holisitic and as involving processes dependent on transformative praxis understood as occurring on all four planes of social being, at up to seven orders of scale and as crucially dependent upon our discourse as a constitutive causally effected and efficacious feature of social life – as discursively inlaid or intricated, that is, as thoroughly conceptually interwoven and indeed moreover as potentially reflexively rearticulated.

Illustrations of the use of dialectical critical realist categories

I now want to illustrate the use of the categorical apparatus of dialectical critical realism by reference to one aspect of the phenomenon of climate change and by reference to the Asian tsunami of December 26, 2004.

We are becoming very aware of the extent to which climate change is the result (largely unintended) of conscious human actions and inactions. Thus using the concept of four-planar social being, we can see, in principle, two-way material transactions with nature as constituting one of the four dimensions in terms of which any social event has to be understood. This four-planar conception presupposes of course that there is an overarching nature in itself. (Critical realism, though susceptible of a humanist rendition, is profoundly anti-anthropocentric.[21]) This defines a notional fifth plane, which constellationally contains the whole of four-planar social being, and from which it is an emergent level.

This four-planar conception gives us a way in which we can think of the effects of particular components in the complex phenomena of climate change, such as, for example, the technology of the motor car and the ecological implications of its use at current levels. Thus we should expect individualism at the plane of social interactions between agents, an oil-orientated foreign policy at the plane of social structure and certain characteristic egocentric patterns of vanity at the plane of the stratification of the embodied personalities of agents to reinforce the effects and consequences for global warming of the use of this technology at the plane of material transactions with nature. The likely effects of such further consequences as desertification, rising sea levels, etc. has been well traced. For our purposes here, the important point is that the development and use of an alternative transport technology has holistic social conditions and implications. Given this, however, global warming and its consequences are something we, in principle, know how to affect, namely by changing a class of human actions.

Let us now turn to a superficially different phenomenon, that of the Asian tsunami. Widely seen and interpreted as a natural disaster, we can observe its horrendous effects on all four planes of social being – through death, destruction of homes and livelihoods, of local communities and societies with their social and material infrastructures, and the psychological traumas of those who witnessed and survived this phenomenon. However, it cannot be regarded as a pure event of nature in which human beings and social policy played no part. This can be shown, for example, by reflection on the fact that there was a substantial time lag between the earthquake caused by the disturbance to the tectonic plates and the effects on the people and land of the regions which were affected. Not only was no warning system, such as existed for the Pacific, in place, but no action was taken by those who did know, such as the American authorities at their base on the island of Diego Garcia or the Thai government, which apparently decided not to sound the alarm out of fear of upsetting the tourist industry. It can even be plausibly claimed that some loss of life could have been avoided if Western tourists and the locals had been more knowledgeable about the behaviour of tsunamis and therefore the need to move to higher ground, rather than to stand and observe the oncoming waves, etc. Here, one can see the consequences in the constitution of this disaster of poor or inadequate geography teaching in schools.

What is striking in the case of the Asian tsunami is the role of inaction or omission, what was not done. This signals the need for a non-monovalent (negativized and dialectical) conception of action and being. More generally, we can list as features of these examples:

- Their global and interconnected character.
- Their holistic and four-planar social laminated character.
- The role played by inaction, omission and absence, as distinct from their positive counterparts, and correspondingly of prevention in pubic policy.

The philosophy of meta-Reality

For the integration necessary to achieve knowledge of an articulated laminated system, we need an integration of the different epistemic perspectives of the various disciplines. That is to say, we need effective cross-disciplinarity, and in order to show how this can be achieved in practice, we need additionally insights from that further development of critical realism, which is the philosophy of meta-Reality. This involves the further deepening of ontology to take in the understanding of being as reflexive or self-conscious, inward and spiritual (5A), the understanding of being as re-enchanted, i.e. as intrinsically valuable and meaningful (6R) and the understanding of being as involving the primacy of identity over difference and unity over antagonism and split, and more succinctly and precisely, as non-dual (7 Z/A). Moreover, it seeks to show how the problems of inter-cultural, interdisciplinary and inter-professional understanding and integration can be resolved in practice.

The main philosophical challenge to the idea that problems of interdisciplinary and inter-professional communication and co-operation can be resolved in practice comes from the thesis of incommensurability. This thesis takes a number of forms, involving scientific incommensurability, moral incommensurability and cultural incommensurability. Elsewhere, I have advanced specific arguments against scientific and moral incommensurability. Here, I want to focus on the question of cultural incommensurability, as manifest in problems of communication and mutual understanding in the different disciplines (and professions), which may be involved in the giving of a laminated explanation of some open-systemic phenomenon or policy recommendation or proposal to ameliorate or transform some complex open-systemic phenomenon in human social life.

The philosophy of meta-Reality formulates two axioms or principles, which may be usefully invoked here.

First, the axiom or principle of universal solidarity (P1) specifies that, in principle, any human being can empathize with and come to understand any other human being.

Second, the axiom of axial rationality (P2) specifies that there is a basic logic of human learning applicable to the practical order, which is accessed by all human communities, irrespective of cultural differentiations.

P2 specifies that there is a decision procedure which allows for the resolution of intercommunal and interpersonal differences, ultimately grounded in the commonalities of our interaction with the natural world. However, even if this were not the case, provided it could be shown that P1 was true, the problem of interdisciplinary understanding would be resolved (what would be lacking would be a means for reaching agreement). I will argue for the necessity of both P1 and P2.

P1 may be motivated by reflection on the contingency of any agent's birth. If they had been born on the same day in a different country or perhaps in the same place, but at a different time, the beliefs, attitudes and habits a person came to adopt would have been very different, Moreover, even if as a result of rational modification of these, they came to these same beliefs, etc., that they held now, they would have come to them by a very different route. They must therefore have had the capacity, which meta-Reality ascribes to their ground-state, to have become very different persons from whom they currently are; and they must therefore also have had the capacity to become one with very different persons from themselves. This capacity to become one with something other than, and radically different from oneself, is of course something that may become stunted in the course of a life, but meta-Reality posits that, however difficult and far removed from current preoccupations, this possibility of becoming one with another remains a permanent and essential possibility.

P2, or the principle of axial rationality, may be motivated by the consideration that people everywhere in the world learn how to raise children, prepare meals, ride bicycles, drive cars, operate computers and machine guns. This learning proceeds by a basic procedure involving the identification and correction of mistakes. As such, it presupposes a universal capacity to learn by a process of correction, and hence (auto-)critique. Since human beings are also linguistic, this capacity must also be expressible in language, and hence within the cultural domain proper, there must also be a mechanism for the identification and correction of mistakes, and hence for critique and reflexive self-critique. Further reflection on this logic of axial rationality at work in our basic material practices of interaction with each other and the world shows that there is always a possibility within any community, say, the members of a discipline or profession, for sustained critical reflection and self-development, and hence, however far it may be removed from something which appears currently feasible, for arriving at an agreement in principle with another community (e.g. discipline or profession) about matters of mutual concern, such as the integration of the knowledge of the different mechanisms in a laminated explanation, which is interacting or intra-acting to produce a particular result!

If the considerations advanced above show that the problem of inter-disciplinary and inter-professional understanding has, in principle, a solution in practice, this may still be very difficult to achieve. And, it is to the resolution of the practical problems involved in interdisciplinary research or inter-professional co-operation that I now turn.

Conditions for successful interdisciplinary research

Before doing so, let me take stock. What is required for successful interdisciplinary research is:

1 Disambiguation of ontology and epistemology;
2 Anti-reductionism;
3 The idea of explanation in terms of a laminated totality;
4 What may be called the holy trinity of interdisciplinary research: meta-theoretical unity, methodological specificity and theoretical pluralism and tolerance;
5 The dissolution of career, administrative and financial barriers to interdisciplinary research.

However, at least two more elements are involved. The first relates to the education of the interdisciplinary research worker. There must be a judicious combination of disciplinarity and interdisciplinarity in their education. Disciplinarity is necessary for the neophyte to get a grasp on the deep structures and mechanisms which constitute the explanatory objects of scientific knowledge and which provide the critical purchase on the potentially ideologically saturated concepts of everyday life and understanding. Without familiarity with the process of retroduction to deep structures which explain phenomenal appearances, the interdisciplinary research worker may stay at a superficial level of understanding of his or her problem.[22] However, without some familiarity with other disciplines and practice at understanding their own vantage points on a common reality, the putative interdisciplinary research worker may revert to mono-disciplinary dogmatism.

Pedagogically, both the case study and the problem method provide a forum in which different disciplines describe and explain in their own way a common pheonomenon, and so may be very useful heuristic devices in the education of an interdisciplinary research worker. But, in addition, they should have familiarity in practising the radical hermeneutic encounter with at least some of the other disciplines that they will be working with. This means that the interdisciplinary research worker should study a second, or even third, discipline as a strong point of reference.

There is a further problem. Research has shown that, whatever the compensating joys of discovery of different and multiple frames of reference for interrogating a given reality,[23] even in the most successful interdisciplinary teams, researchers may suffer from a mild but definite form of alienation which has been described as a feeling of 'stray'.[24] This comes from not having a sense of a secure recognized place or home in a single disciplinary tradition. This too can, to some extent, be compensated by their developing roots in other disciplinary traditions too, and perhaps also by greater career flexibility so that periods of interdisciplinary research work can be punctuated by periods of working on the problems of their home discipline.

A radical objection to the idea of interdisciplinary research work must be met. It may be pointed out that, in the history of science, interdisciplinary research, when successful, tends to generate or constitute a new disciplinary research tradition. However, the fact that a laminated totality has stabilized and cohered to constitute a discipline in its own right is no objection to interdisciplinary research work. For such a new discipline, such as, for example, ecology, must participate in interdisciplinary research projects, constructing new laminated totalities, with other, already constituted disciplines. In other words, the fact that there is an institutional, as well as an individual, dialectic of disciplinarity and interdisciplinarity does not vitiate the importance of interdisciplinary work.

Of all the practical problems involved in interdisciplinary research, perhaps the biggest is the old natural science/social science divide,[25] evidently a legacy of C.P. Snow's 'two cultures'. In order to counter this, it could perhaps be suggested that a subject from the other side of the divide should be one of the subsidiary disciplines in the education of the interdisciplinary research worker.

Elsewhere, I have elaborated an ideal or canonical form for a typical critical realist applied research project.[26] This will include work in the applied open-systemic context at two margins of inquiry – an intensive margin, in which wholes are packed into their parts and an extensive margin, in which parts are spread out over their wholes.[27]

Forms of critique

So far in this chapter I have developed the concepts necessary for the understanding of complex phenomena such as climate change, and for the reconstruction of contemporary discourse on climate change, but we need of course at the same time to critique existing actualist, reductionist and monodisciplinary accounts of such phenomena.

In general, critique will take a triple form. It will involve in the first place *immanent critique*, that is taking a system of thought on in its own terms, showing how it involves various internal contradictions and aporiai. This process of immanent critique may be radicalized through various forms of transcendental and dialectical refutation to the point which involves what I have called an Achilles heel critique, that is a critique of a system of thought on the very point where it is believed to be, and believes itself to be, strongest – such as the Achilles heel critique of empiricism on the grounds of its incapacity to sustain coherent concepts of experience, especially experimental activity, in science.

The second major form or level of critique is that of omissive critique or metacritique$_1$. This involves the elucidation of the generative absences at work in the system of thought, such as the absence of disambiguated ontology and the non-actual real in empiricism. This level of critique depends of course on the rectification of the identified absences in a more comprehensive and coherent account, which is the essence of the progressive dialectical movement of thought.

The third form or level of critique is that of an explanatory critique or metacritique$_2$. This involves a substantive explanation of not only what is wrong or inadequate in a system of thought, but why it was believed, that is (considering different modalities of this explanatory form), how it came to be generated, accepted and reproduced. Such a form of critique will of course inevitably pass over to a critique of the objects generating the inadequate, misleading, or superficial consciousness. It may be further extended to show the full range of the baneful effects of the faulty system of belief, and its causes.

Critique of inadequate metatheories

Perhaps the biggest obstacle to successful interdisciplinary research work, and therefore to the understanding of complex open-systemic phenomena such as climate change, lies in the way in which woefully inadequate metatheories and methodologies continue to inform the practices of the various disciplines which continue to seek to understand such phenomena in an actualist, reductionist and often still fundamentally mono-disciplinary way. Elsewhere, Berth Danermark and I[28] have described the effects of (often unconscious and set in the context of a 'TINA compromise form'[29]) empiricism, neo-Kantianism, superidealism, hyper-hermeneutics, strong social constructionism, poststructuralism, postmodernism, etc. The dominance within these traditions of the epistemic fallacy, actualism and monovalence undermines the possibility of thinking the concepts necessary for the understanding of complex open-systemic phenomena, as it also undermines the possibility of a successful resolution of the problems of interdisciplinary and interprofessional communication and understanding in practice.

The full development of critical realism through dialectical critical realism and the philosophy of meta-Reality also allows a *generalized critique of reductionism* to include, for example, besides the effects of actualism at 1M, those of mono-valence, de-negativization and de-processualization at 2E, extensionalism, detotalization and decontextualization at 3L and de-agentification including voluntarism and reification at 4D.

Concrete utopianism

Having shown how the development of critical realism allows a reconstruction of the concepts necessary for understanding complex open-systemic phenomena such as climate change and for critiquing inadequate accounts of such phenomena, the question arises as to how we can use such knowledge to change the world. The full development of the theory of explanatory critique understands it as involving a complex of explanatory critique, what I have called concrete utopianism and a theory of transition, in dialectical unity with an emancipatory axiology of transformative practice. In this ensemble, concrete utopianism plays a crucial role. It involves thinking how a situation or the world could be otherwise, with a change in the use of a given set of resources or with a different way of acting subject to certain constraints. This mode of thinking forms the basis of an ethics oriented to change, in which we think alternatives to what is actualized on the basis of given possibilities, possibilities which were actualized in one way but could be (or might have been) redeployed or actualized in another.

Traditional leftist critiques of utopianism have actualistically failed to notice that what is, is only one possible world and that it, moreover, always presupposes the possibility of other worlds. Radical intellectuals need to show in detail how alternative futures can be coherently grounded in the deep structures of what already exists, of what people already know and have. Without this exercise, they will not be able to make out a persuasive case for change. With it, there may yet be a way in which, combining realism (not, contra Gramsci, pessimism) of the intellect with optimism of the will, humanity can usher in that future of which the youthful Marx said, 'The world has long since dreamed of something of which it needs only to become conscious for it to possess in reality'.[30]

Notes

1 See Roy Bhaskar and Berth Danermark, *Being, Interdisciplinarity and Well-Being: A Study in Applied Critical Realism*, forthcoming, Routledge, London and New York, 2009. For the status of the argument in terms of the general architectonic of critical realism, see also Chapter 9 of my forthcoming book with Mervyn Hartwig, *The Formation of Critical Realism: A Personal Perspectrive*, Routledge, London and New York, 2009.

2 See my *A Realist Theory of Science*, 1975/2008, 4th edn, Routledge, London and New York, Appendix 2, Chapter 2.

3 Especially in my books *A Realist Theory of Science*; *The Possibility of Naturalism*, 1979/1998, 3rd edn, Routledge, London and New York; and *Scientific Realism and Human Emancipation*, 1986/2009, 2nd edn, Routledge, London and New York.

4 This term of art derived originally from Andrew Collier, *Scientific Realism and Socialist Thought*, Pluto, London, 1989.

5 R. Bhaskar and B. Danermark, 'Metatheory, interdisciplinarity and disability research: a critical realist perspective', *Scandinavian Journal of Disability Research*, Vol. 8, No. 4, 2006, pp. 278–297.

6 K.G. Høyer and P. Næss, 'Interdisciplinarity, ecology and scientific theory', *Journal of Critical Realism*, 7.2, 2008.

7 G. Brown, 'The ontological turn in education', *Journal of Critical Realism*, 8.1, 2009.

8 See *The Possibility of Naturalism*, p. 59.

9 See my *Dialectic: The Pulse of Freedom*, 1993/2008, 2nd edn, Routledge, London and New York.

10 See *Dialectic*, pp. 129 ff.

11 This has been pointed out by George Steinmetz in an illuminating article — see G. Steinmetz, Critical realism and historical sociology. A review article, *Comparative Studies in Society and History*, Vol. 40, 1998, pp. 170–186.

12 See *Scientific Realism and Human Emancipation*, p. 109, and *Dialectic*, p. 123.

13 See *Dialectic*, p. 123.

14 Björn Blom and Stefan Morén have formulate a 'CAIMO' model for social intervention practice (which can be extended to policy generally), where C signifies context, A action, I intervention, M mechanism and O outcome – see 'Explaining human change: on generative mechanisms in social work practice', *Journal of Critical Realism*, 2(1), 2003.

15 See Bhaskar and Danermark, 2006, p. 293.

16 See *Scientific Realism and Human Emancipation*, pp. 106 ff.

17 See *A Realist Theory of Science*, pp. 121–122.

18 See *A Realist Theory of Science*, p. 70.
19 See Jenneth Parker, 'Situating education for sustainability: a framework approach', *Journeys around Education for Sustainability*, ed. J. Parker and R. Wade, London South Bank University, London, 2008. Cf. also Peter Dickens, *Reconstructing Nature*, Routledge, London, 1996.
20 R.M. Hare, 'Meaning and speech acts', *Philosophical Review*, 1970, pp. 19ff.
21 See S. MinGyu, Bhaskar's Philosophy as Anti-Anthropism, *Journal of Critical Realism*, 7(1), 2008.
22 See Leesa Wheelahan, *Why Knowledge Matters in Curriculum*, Routledge, London and New York, forthcoming 2010.
23 M. Nissani, 'Ten cheers for interdisciplinarity: the case for interdisciplinary knowledge and research', *The Social Science Journal*, 34(2), 1997, pp. 201–216.
24 See the report on the Swedish Institute of Disability Studies at Orebro described in *Being, Interdisciplinarity and Well-Being*, Chapter 4.
25 Sarah Cornell has recently reported this as a principal practical barrier to successful interdisciplinary cooperation on climate change – compare Sarah Cornell, Chapter 7, this volume.
26 See *Being, Interdisciplinarity and Well-Being*, Chapter 5.
27 See *Scientific Realism and Human Emancipation*, pp. 111–112; and *Dialectic*, pp. 125–126.
28 *Being, Interdisciplinarity and Well-Being*, Chapter 5.
29 See *Dialectic*, pp. 116–119.
30 Marx to Ruge, September 1843, *Early Writings*, Harmondsworth, 1975, p. 209.

2 Critical realist interdisciplinarity

A research agenda to support action on global warming

Sarah Cornell and Jenneth Parker

This chapter will pick up on some themes from Roy Bhaskar's previous chapter and discuss them in relation to the tasks of developing an adequate range of knowledge about climate change and developing effective interdisciplinary agendas. Our position is that knowledge for climate change, and for sustainable responses to it, must involve a wide disciplinary range from biophysical climate science, through to social science understanding of social structures, including the cultural and ethical aspects that frame and motivate human action. In this respect we situate this discussion within a global systems approach, expressed in Table 2.1 below.

This holistic perspective indicates that we cannot simply apply philosophical perspectives to vindicate and clarify existing science. The ontology of a joined-up world in process expressed in Bhaskar's Chapter 1 (this volume) is a starting point from which critique of reductive disciplinary tendencies is inevitable. This ontology underpins our discussion. This starting point also dictates that consideration of any one area of knowledge involves an interdisciplinary understanding of the contexts of development and application of the wider field. We focus here on climate science, but we also want to ask how critical realist tools and approaches can help us link this science with other, wider areas of knowledge and research.[1] We aim to draw out issues from climate change knowledge for discussion and investigation, laying the ground for the further development of a research agenda, to which other chapters also contribute in different ways.

Climate science

Changes in climate are dynamically related to changes in a variety of linked Earth systems – atmospheric chemistry; soil loss and degradation; deforestation and other major changes in land use; biological changes in our oceans and so on (Steffen *et al.*, 2004; Bretherton, 1988, Figure 1). These multiple and interconnected linkages, which mean that local and transient perturbations can have global consequences, have become the focus of worldwide scientific research efforts. Yet the concept of the 'Earth system' (as with all concepts) has developed with cultural and metaphysical baggage. 'Gaia', the theory that addresses the complex interactions between Earth's living and non-living components, can

Table 2.1 Current global system concerns (some indicative examples using a basic critical realist laminated system framework)

Questions crossing over two domains	*Questions linking adjacent domains*	*Questions crossing over three or more domains*
	Cultural domain	
The contribution of Sci-Art	Framing of contemporary models and scenarios	
	Values regarding future generations and our future selves	Urbanisation
Knowledge needs for governance/participation	Visions for futures	Resilience/vulnerability
	Socio-institutional domain	
	Joined up policy	
	Climate agreements	
	Bio-diversity agreements	Education for response to climate change
Livelihoods and migration	Legal and economic frameworks	
Eco-social history/futures	**Socio-ecological domain**	
	Low energy technologies	Integrated environmental resource management
	Alternative energy and fuel sources	
	Low greenhouse gas energy	
Indexing the eco-efficiency of new/proposed technologies	**Ecological domain**	Governance and justice in ecological era
	Models of ecosystems	
	Bio-physical life support systems	
	Ecological regeneration/restoration	Security/risk
	Atmospheric chemistry	
	Cosmological/atomic/chemical domain	

raise as many barriers as it can provide useful metaphors. A change to a more philosophically informed approach that does not appear to dictate a mythological or teleological dimension, such as critical realism can provide, will be welcome to many for these reasons alone.

The critical realist ontology of a differentiated and dynamically relational world, as outlined in the previous chapter, can support the importance of climate science and vindicate its increasingly interdisciplinary development. Why might this be needed? We are just emerging from a situation where sciences claiming more certain knowledge are privileged over those sciences working in more complex, less predictive (or predictable) terrain. Climate science has its deepest roots in physics; it is laden with presumptions both within and outside the research community that its investigations should lead to greater certainty and predictive power (in the sense

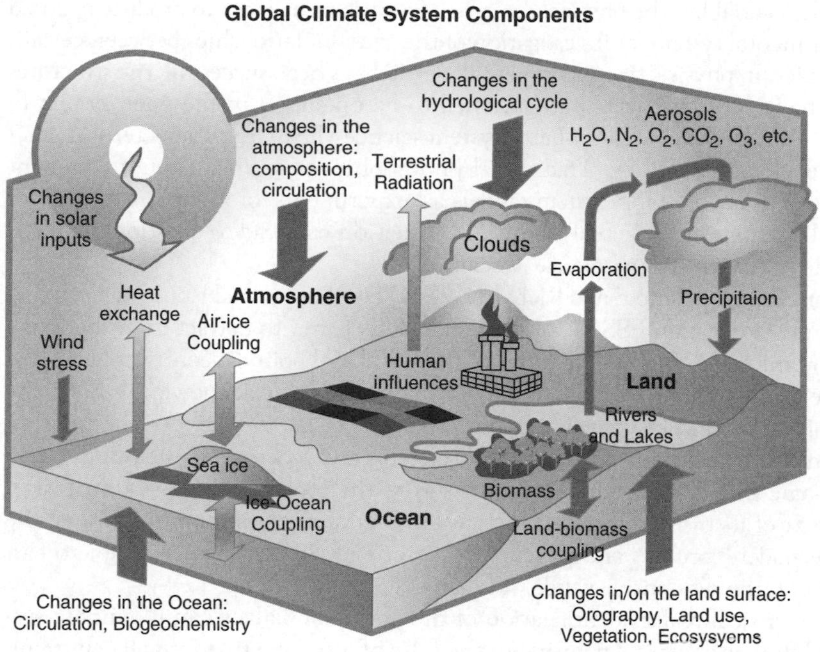

Figure 2.1 The global climate system (from NSF website http://www.nsf.gov/geo/adgeo/geo2000/ideas_society.jsp;).

of generating new testable hypotheses). This is why climate sceptics focus in on 'weak points' where there is less certainty or determinacy. But the pressing questions of climate change are increasingly being oriented towards the dynamic, often co-adaptive, interrelations of life and its physical environment. Here, interdisciplinarity is vital, and a clear articulation is needed of why this might be so, and why it does not represent a dilution or weakening of knowledge. It is here that the critical power of the critical realist explanation of the epistemic fallacy (Chapter 1, p. 1) becomes crucial. For example, the epistemic fallacy contributes to the notion that the structures and entities with which ecology has to deal are in some way less real, and hence less important, because they cannot be so certainly demonstrated. A critical realist ontology can combine ontological realism – the objective reality of the life processes on the planet – with a recognition of the necessary complexity and difficulty of prediction of phenomena in open systems (Chapter 1, p. 3). Climate change underlines that 'ecology specifies the conditions of possibility of human material practices' (Chapter 1, p. 12), and the fact that we humans can radically alter planetary ecology, to the extent of fatally undermining our own life-support systems.

Traditional philosophies of science have taken physics as their exemplar case. They have been slow to update themselves in the light of interdisciplinary developments. While the world of actual phenomena is an open system, with limitless variables, the physical scientist can work to attempt to produce a closed experimental system for investigation of the causal relationships between certain variables. In physics, this experimental method has been successful; the structures that have been discovered are assumed to be operating in the open system of the world. In the context of Earth system science (ESS), the characteristic practice is global modelling. Traditional philosophies of science cannot comment constructively on Earth system models as 'experiments' or as representations or simplifications of the actual world. This situation can lead to questioning of the validity of Earth system science as science.

For example, Clifford and Richards (2005) cite Popper and Rouse as providing grounds to question ESS as valid science. This is not to detract from the many important issues they raise in terms of the social and political context of ESS and its relations with power. However, crucially, CR can acknowledge both the transitive and intransitive dimensions of science (Chapter 1, p. 1) – the fact that it can reasonably claim to be about real systems and processes in the world, but at the same time is susceptible of analysis as to the social, political and historical context of its own productive practices. It is vital to clarify that to critique the power-laden contexts and policy applications of climate science is not to cast doubt on the substantive validity of many of its outputs.

The critical realist identification of the epistemic fallacy can illuminate sustainability questions by pointing to the folly of assuming that formally epistemically convenient forms of unitary measurement are the best option for describing a complex world. Climate science has developed from the investigation of manageable areas for which we have the possibility of providing sufficient data within a relatively short time frame. In the social world particularly this results in a 'Procrustean bed'[2] where ecological and social reality is distorted to fit inadequate tools of measurement. CR can also allow us to acknowledge the historically conditioned data with which climate science has to work – satellite data can tell you about land use in some respects, but not about the social and political contexts and causes that frame human activities in different areas of the world. With regard to sustainability, climate science can fall foul of the presupposition of the additivity of knowledge, and the CR recognition of scales and of emergence offers an important check. The 'emergent intermeshing of the different mechanisms . . . require genuinely synthetic interdisciplinary work, involving the epistemic integration of the knowledges of the different mechanisms' (Bhaskar, Chapter 1, pp. 4–5).

Critical realist interdisciplinarity applied to Earth system science

Here we want to explicitly focus on the intellectual opportunities of extending from climate to the global socio-ecological system, even though most ESS does not generally proceed so far as yet. In attempting to study the Earth system as a whole phenomenon, the research object is an already intensively relational set of systems and sub-systems. Within this mix it is recognised that we can identify certain key sub-systems with their own laws and tendencies. The overall research question is the effect on the whole system of changes within certain specific sub-systems – and, given the current concerns about anthropogenic climate change, attention is mostly focused on the impacts of changes brought about by human practices, ranging from production, consumption, transportation, settlement, to resource use and so on. The Earth system research programme is evolving into one that is attempting to quantify and categorise effects on our life-support systems in ways that can support effective decision-making for future human sustainability (see, for instance, the Millennium Ecosystem Assessment, 2005).

We can expect to be able to develop some correlations between changes in these sub-systems with a high degree of certainty. For instance, knowledge is robust about the links between increased industrialisation and the changes due to higher greenhouse gas concentrations in the radiative forcing of the Earth's energy balance. However, other correlations are much weaker, as we are working with the ultimate joined-up system. A pressing contemporary question for research and policy is that of the 'climate sensitivity' – how much warmer will the world be if carbon dioxide concentrations double? There is no doubt in the international, interdisciplinary research community that the world will be warmer – not cooler, nor even the same – but the sensitivity of our planetary thermostat is still debated.[3] This number matters enormously to those wanting to project future climate. Even without a consideration of what society might choose to do to 'fix' climate in future, this climate sensitivity number is proving hard to pin down with the wished-for precision, involving as it does complex adaptations in the multiple biogeochemical and ecological sub-systems.

When we attempt to model complex Earth system processes, it becomes difficult to differentiate between studying the sub-systems interacting and/or studying the interactions themselves. As Bhaskar puts it, 'holistic causality depends upon internal relationality' (Chapter 1, p. 7). This challenge is most evident in the context of the deep interdisciplinarity of the human dimensions of global environmental change: are individuals and institutions best thought of as a social sub-system, or as an 'external forcing' of the Earth system? This question echoes Heisenberg's 'Is it a wave or a particle?' In computer modelling of the Earth system, the criteria of success are the fit of the model to observations of actual Earth processes and outcomes over time. This then underwrites the validity of predictions made on the basis of the model. Predicting human systems is proving contentious, unsurprisingly. We argue that the complex, systemic nature of the climate research task can be brought into higher critical definition through explicating it in the context of CR approaches. Here the problem is not the aggregation of disciplines (that is, a problem in the working practices of knowledge production), it is rather the disaggregation of holistic phenomena into manageable areas. CR can help to pinpoint the ontological and epistemological issues when undertaking such a task (see Chapter 1, p. 3).

The CR account of the ontology of a joined-up world can help us to question the implicit ontology of ESS to date. Arguably, the development of ESS has been an accretion of discretely identified elements, and the ontological breakthrough that is needed has been indefinitely postponed. This lack of ontological renewal of the basic presuppositions of the science is likely to be a major factor in the difficulties ESS now confronts in linking with the human sciences. In his film, *An Inconvenient Truth*, Al Gore notes that humanity often finds it difficult to 'join up the dots'. In respect of ESS, we are not taking on board the crucial CR point that joining up the dots changes the nature of the dots. In CR terms, 'When the mechanisms themselves change, and are thus emergent, we can talk of intradisciplinarity'. . . (Chapter 1, p. 5). Thus a very important aspect of interdisciplinary engagement is that this engagement itself may, indeed probably will, change the sciences. Much interdisciplinary 'engagement' is more like the recreation of certain aspects of the 'participating' sciences in order to equip them adequately to answer the research question. This raises thorny issues of the relative perspectives framing questions from 'within' different sciences and the assessing of the place of those questions within the wider research programme.

We provide an overview of the nature of the research task informed by the dialectical critical realist framework outlined in Figure 2.1 above. An admittedly limited schematic presentation such as this begins to provide an appreciation of the interlinked nature of the knowledge agenda and sub-agendas prompted by climate change and can help us to better conceptualise both the knowledge inputs and practices that are required and their essential relationships.

Implications of CR interdisciplinarity for responses to global warming

Can we identify pointers from an ontology of a joined-up world that can help us to focus our efforts to respond to global warming more effectively?

Radical inadequacy

We can certainly use frameworks derived from CR and a more interdisciplinary ESS to identify the radical inadequacy of any proposed 'solutions' that do not take account of the connective aspects of the socio-ecological system. Even though ESS is inevitably tending towards more integrative interdisciplinarity, many 'solutions' being mooted have not been developed with even a rudimentary awareness of the connectedness of the socio-ecological system (Young *et al.*, 2006; Norberg and Cumming, 2008). The critical realist elaboration of the laminated systems understanding (Chapter 1, p. 5) can assist by providing some guidelines here.

Regarding the landscape of related areas above, the epistemic fallacy is still widely operative in that there is an overwhelming tendency to concentrate on areas of action that we have information about, and ignore those areas where we are relatively ignorant. Critical realism continually reminds us of the enormity of the unknown and unproven: the world does not correspond to our knowledge of it. These concepts support the case for the precautionary principle (Parker, 2001) in relation to the risks of human intervention in ecological systems. We still face a huge gap in knowledge regarding how to assess and optimise socio-ecological changes. These knowledge needs should be assessed and priorities for the development of new knowledge assigned accordingly. Pulselli *et al.* (2008) have proposed an approach to studying whole system behaviour and the interactions between natural and human agents in a way that addresses the world's real limits. The developing fields of ecological economics (Costanza *et al.*, 1997) and resilience (Walker and Salt, 2006) are also focused on the nature and sustainability of human environmental interactions

This blindness to the poorly known combines with a physical world mindset (and 'carbon reductionism', see Chapter 3 this volume) to produce radically inadequate solutions that can set in train social responses and crises that can worsen the situation. In the contemporary climate context, most people are familiar with the potential for maladaptive efforts to increase bio-fuels. Even the useful approach for stabilising fossil fuel emissions of Pacala and Socolow (2004) can be criticised on this count. Whilst their 'wedge' concept is a very useful heuristic device, and has brought home to many the importance and feasibility of a multi-headed effort to respond to climate change rather than a 'magic bullet', each wedge is itself in need of interdisciplinary unpacking. For example, fuel substitution, one of the wedge options, relies heavily on technical developments that do not recognise the joined-up socio-ecological system on which their success hinges. The wedge concept also gives the idea that several things can be optimised independently, yet the wedges overlap and interact and should be subject to systemic analysis.

There are risks too in the over-hasty synthesis of existing knowledge; we need to keep re-stating the truism that not everything interdisciplinary is good or useful. The tacit axioms of one knowledge field engaged in an interdisciplinary venture could be perniciously undermining the whole project. As already mentioned, in the socio-ecological context, there can be a 'strong entwinement' of the ontological and epistemological aspects of conceptualisation of processes and their uncertainty (Dequech, 2004). The process of interdisciplinary knowledge synthesis needs to be subject to critical tools and processes that can be informed by CR.

Bridging nature, society and culture

Critical realism provides a structure that relates the natural and social worlds in just the way that analyses of sustainability require. To explain climate change, we need an epistemology that recognises social dimensions of knowledge, but also an ontology that asserts the reality of the material dimension of the problems. Critical realism can help achieve a structured overview by virtue of its descriptions of schematic relations between different 'levels' of emergent reality and this can help prevent the complex interdisciplinary areas of study necessary for sustainability from becoming a chaotic mix of social interpretation and physical facts.

The philosophy of critical naturalism (expounded in Chapter 1, p. 8) demonstrates that critical realism can coherently combine assertion of the independently existing powers and capacities of natural systems with the capacities of humans to take transformative action. One key aim of critical realist analysis of social science was to assert the reality of social structures. In opposition to those who argue that there is no such thing as 'society', critical realism argues for the reality and causal powers of social structures and hence for the importance of analysing and understanding them for human intentional action. Because critical realism is a non-reductive philosophy, supporting the reality of structures, entities and powers at different levels of complexity in the world, it argues against the reduction of social reality to mere amalgams of individual motivations, powers or tendencies. The CR recognition of four-planar social being and the causal powers of social systems and institutions (Chapter 1, p. 9) is essential to inform strategies for action that can combine transformative action across the personal, social relational and social structural spheres and scales.

Critical realism can accommodate valuable post-modern emphases on cultural interpretation within a recognition of the independent reality of the world. Post-modern philosophy, on the contrary, has paid a great deal of attention to questions of interpretation, to the neglect of material questions. We need to give the necessary importance to questions of interpretation and meaning within social critique whilst maintaining the reality of the world external to human ideas. Social movement theory and practice has recognised the role of culture in activism and the transformation of practice. As Bhaskar puts it, 'transformative praxis . . . [is]. . . crucially dependent upon our discourse as a constitutive causally effected and efficacious feature of social life' (Chapter 1, p. 16). In this way the critical realist perspective is essential for a critical cultural theory that can engage with questions of power (Agger, 1992). Further critical realist engagement with socially critical cultural studies is clearly desirable, as argued by Cheryl Frank in Chapter 6 of this volume.

Scalar information and causal explanation

There are many challenges in modelling at different scales. For humans and ecosystems, *where* a cloud forms matters at least as much as that it forms at all, but climate impacts modelling grapples with two challenges at this more localised scale, where typically a simplified climate model (or a statistical treatment of downscaled global model output) is coupled to a econometric model. First, the predictive power of global climate models, which capture bulk properties well, is weaker at regional to local scales, which are more affected by the conceptual abstractions from the actual climate processes (parameterisations) that make the model manageable and fit for its primary purpose of understanding global processes. Furthermore, many important insights into the consequences of, say, a cloud's formation are not captured by the coupled tools of climate physics and economics; adding on more model components may still fail to provide knowledge of real human (and ecological) experience.

In terms of assigning causal explanation to systemic phenomena, CR emphasises the necessity of exploring the conjunctive multiplicity of causally operative systems operating at different scales. We cannot arrive at ways forward without performing some kinds of basic scalar checks on our explanations, particularly with respect to more local human possibilities for action within a globalised economy. Localised issues such as food crises raise concerns of the relative power of local economic decision-making as against the power of global capital (Lacey and Lacey, Chapter 11, this volume). More literacy about the scalar effects of complicating interactive factors is also needed in order that policy makers and the wider public can appreciate the scope and effectiveness of climate science. The question of explanatory power at different scales also has a strong relationship to whose questions science is attempting to answer. As Hornborg has argued:

> As long as the primary knowledge interest of a science is to generate growth strategies for individual companies or nations, it is only natural that its fundamental assumptions should differ from those required of a science of global resource management.
>
> (2003, p. 214)

Research agendas for CR interdisciplinarity: towards an enabling philosophy for interdisciplinary science?

Increasingly, what justifies science (and the vast expenditure on it) is that it may help answer questions set by a joined-up world. If this is so, then the adequacy of science is partly determined by our ontology (or lack of one) of a joined-up world. Especially in the context of a 'planet under pressure' because of unbalanced human–environment interactions (Steffen *et al.*, 2004; Millennium Ecosystem Asssessment, 2005), we cannot afford to throw money at research programmes that fail to progress towards concrete knowledge and conjectures. Interdisciplinary trial and error are proving costly in terms of time and resources. More attention to philosophy and greater literacy in methodological issues of interdisciplinarity may help us to focus our research and change projects more effectively. What kind of rationale can be supported by CRI? Does not complexity and interdisciplinary science require a more actively engaged philosophy of (or for) science? We suggest that CR can be deployed here, in developing further enquiry into the dialectical nature of the theory/practice relationship in the original CR transcen-dental deduction, and in developing more focussed tools and approaches for assessing interdisciplinary research projects, methodology and interdisciplinary participants.

Notes

1 This chapter also draws on Sarah Cornell's discussion of climate science in Chapter 7 of this volume.

2 Procrustes was a robber in Greek mythology who fitted travellers into his bed by stretching or lopping off their limbs.
3 The best estimate, from models and observational data, is a 3°C global average temperature increase, ranging from 2 to 4.5°C (IPCC, 2007).

References

Agger, B. (1992). *Cultural Studies as Critical Theory*. London: Falmer Press.

Bretherton, F. (1988). *Earth System Sciences: A Closer View*. Washington DC, Earth System Sciences Committee, NASA.

Clifford, N. and Richards, K. (2005). Earth system science: an oxymoron? *Earth Surface Processes and Landforms*, **30**, 379–383.

Costanza, R., Cumberland, J. C., Daly, H. E., Goodland, R. and Norgaard, R. (1997). *An Introduction to Ecological Economics*. Boca Raton, St. Lucie Press.

Dequech, D. (2004). Uncertainty: individuals, institutions and technology. *Cambridge Journal of Economics*, **28**, 365–378.

Hornborg, A. (2003). Cornucopia or zero-sum game? The epistemology of sustainability. *Journal of World Systems Research*, **IX**(2), Summer.

Millennium Ecosystem Assessment (2005). *Living Beyond our Means: Natural Assets and Human Well-being*. Statement of the MA Board, March.

Norberg J. and Cumming, G.S. (eds.) (2008). *Complexity Theory for a Sustainable Future*. New York: Columbia University Press.

Pacala, S. and Socolow, R. (2004). Stabilization wedges: solving the climate problem for the next 50 years with current technologies. *Science*, **305**, 968–972.

Parker, J. (2001). The precautionary principle, in Chadwick, R. (ed.) *The Concise Encyclopedia of the Ethics of New Technologies*. London: Academic Press.

Pulselli, R.M., Pulselli, F.M. and Rustici, M. (2008). Emergy accounting of the Province of Siena: towards a thermodynamic geography for regional studies. *Journal of Environmental Management*, **86**, 342–353.

Steffen, W., Sanderson, A., Tyson, P.D. *et al.* (2004). *Global Change and the Earth System: A Planet Under Pressure*. New York: Springer.

Walker, B.H. and Salt, D.A. (2006). *Resilience Thinking: Sustaining Ecosystems and People in a Changing World*. Washington DC: Island Press.

Young, O., Berkhout, F., Gallopin, G., Janssen, M., Ostrom, E. and van der Leeuw, S. (2006). The globalization of socio-ecological systems: an agenda for scientific research. *Global Environmental Change*, **16**(3), 304–316.

3 Seven theses on CO_2-reductionism and its interdisciplinary counteraction

Karl Georg Høyer

Introduction

The point of departure is the claim that the contemporary discourse on climate change is dominated by CO_2-reductionism. This is the form of reductionism where complex phenomena interconnecting both nature and society are reduced to one singular issue: emissions of CO_2. Implications are that mitigation efforts are delimited to a matter of reductions in just these emissions, entailing also a focus only on technological and simple economical solutions to achieve such reductions. Many examples can be given. For one, the Norwegian Low Emission Committee in recommending mitigation measures only focused on CO_2, and only came up with technological solutions to attain a low emission society (Randers *et al.*, 2006). The Committee chairman (Randers, 2006) claimed that about a 70 per cent reduction in CO_2 emissions could be achieved quite easily without any major societal changes, a claim that can only be substantiated within a rather narrow reductionist position.

The contemporary discourse is largely based on three *substantively* different but interrelated types of reductionism:

1. Climate gas issues are reduced to CO_2
2. Energy issues are reduced to CO_2
3. Environmental issues are reduced to CO_2.

Elaborations on all three are made in more detail throughout the chapter. They are dominant both in political and science discourses. In the case of science, this dominance, however, is mostly related to the particular field of research focusing on mitigation measures, whether positioned within social, economical, natural or technological sciences. In the case of politics, it is more extensive, not the least enforced by the unilateral focus on combating CO_2 emissions by highly influential environmental organisations. In Norway two such organisations are solely focusing on delimited technological solutions, even explicitly denying that broader societal issues are challenged. Thus both institutionally and economically they have become 'married' to the development of separate technological solutions: one on carbon – actually CO_2 – capture and storage (CCS) and the other on hydrogen fuel in transportation. Politically more controversial issues of economic growth or even growth in energy consumption are well hidden beneath the carpet (Høyer, 2007).

With the historical roots of the environmental movement in mind, this raises the question whether these new organisations should any longer be considered to belong to this movement. Since its modern advent in the early 1960s, true anti-reductionism, and the related interdisciplinarity, has been at the very heart of the environmental movement, not only as a condition for understanding environmental problems and their causes, but also for approaching possible solutions. Actually the environmental movement and discourse have played crucial roles in continually raising critical issues about the need for interdisciplinarity in science. Historically, the development of the Intergovernmental Panel on Climate Change – IPCC – may even be considered as a result and a true heritage of these processes (Høyer, 2007; Høyer and Næss, 2008). And when I am focusing on CO_2-reductionism here, this does not apply to the IPCC. Through all our 40-year history of environmental science and politics, this panel is the most comprehensive, counter-reductionist and interdisciplinary scientific endeavour ever achieved, clearly emphasised by its encompassing international structure and co-operation.

In this chapter I shall address the various forms of reductionism through *seven theses*. I shall discuss counteractions necessary to move societies towards the *wholeness* required, where CO_2 is reunited with the other factors from which it has become disconnected during the last two decades. This is done step by step, from the first more delimited theses to the last more comprehensive ones. These seven theses are:

1 Reuniting CO_2 with other greenhouse gases
2 Reuniting CO_2 with fossil energy
3 Reuniting CO_2 with energy
4 Reuniting CO_2 with consumption
5 Reuniting CO_2 with economic growth
6 Reuniting CO_2 with sustainable development
7 Uniting CO_2 with the post-carbon society.

Reuniting CO_2 with other greenhouse gases

Greenhouse gases are a multitude of both natural and anthropogenic gaseous constituents of the atmosphere. They are primarily water vapour (H_2O), carbon dioxide (CO_2), nitrous oxide (N_2O), methane (CH_4) and ozone (O_3). However, there are an additional number of entirely human-made greenhouse gases in the atmosphere, such as the halocarbons and other chlorine- and bromine-containing substances, dealt with under the Montreal Protocol to combat depletion of stratospheric ozone. Besides CO_2, N_2O and CH_4, the Kyoto Protocol on combating climate change deals with the greenhouse gases sulphur hexafluoride (SF_6), hydrofluorocarbons (HFCs) and perfluorocarbons (PFCs). Within the last two groups, there are a large number of individual gases with different properties, atmospheric lifetimes and global warming potentials (IPCC, 2001, Annex B). Just within the HFCs there are more than ten individual gases with global warming potentials varying from a 100 to 12,000 in a 100-year time horizon, as adopted by the IPCC (Solomon *et al.*, 2007).

Table 3.1 gives an overview of the properties of the currently most important anthropogenic greenhouse gases.

According to IPCC (AR4, 2007) since pre-industrial times, the concentration of atmospheric CO_2 has increased by 36 per cent, from the pre-industrial value of about 280 ppm to 380 ppm in 2005. The annual CO_2 concentration growth rate was larger during the last ten years (1995–2005) than it has been since the continuous atmospheric measurements began in around 1960. Main sources of the increased concentrations are emissions from use of fossil energy and from the effects of land use changes on plant and soil carbon, including deforestation. From 1750 it is estimated that about two-thirds of the anthropogenic CO_2 emissions stem from fossil energy use and the other third from land use changes. Not only the concentration, but also the emissions of CO_2 have continued to increase over the last few decades, from fossil energy sources with a yearly average of about 6.4 GtC in the 1990s to about 7.2 GtC during the years 2000–2005 (1 GtC = 3.67 $GtCO_2$).

CO_2 may have relatively long residence times in the global climate gas system. If net global anthropogenic emissions were kept at current levels, this would lead to a nearly constant rate of increase in atmospheric concentrations for at least two further centuries, reaching about 500 ppm, almost twice the pre-industrial concentration by the end of the twenty-first century. Carbon cycle models show that stabilisation of the concentration of CO_2 at its present level could only be achieved through an immediate reduction of emissions by 50–70 per cent and further reductions thereafter.

Table 3.1 Atmospheric lifetimes and global warming potentials of important greenhouse gases

Greenhouse gas		*Atmospheric lifetime*	*Global warming potential (GWP) for three different time horizons*		
Gas name	*Chemical symbol*	*Lifetime in years*	*20 years*	*100 years*	*500 years*
Carbon dioxide	CO_2	Variable	1	1	1
Methane	CH_4	12	72 (x)	25 (x)	7.6 (x)
Nitrous oxide – laughing gas	N_2O	114	289	298	153
Perfluor carbons	e.g. CF_4	50,000	5210	7390	11,200
	e.g. C_2F_6	10,000	8630	12,200	18,200
Sulphur hexafluoride	SF_6	3200	16,300	22,800	32,600
Hydrofluorocarbons	e.g. HFC-134a	14	3830	1430	435

x = GWP for methane includes indirect effects from enhancement of tropospheric ozone and stratospheric water vapour.

Adapted from Solomon *et al.* (2007).

Of all greenhouse gases the global warming effect of methane – CH_4 – is second in strength only to CO_2. According to the IPCC fourth assessments (AR4, 2007), current atmospheric concentration of methane averages about 1770 ppb (parts per billion), representing an increase by about 150 per cent since pre-industrial times in about 1750. The last 15 years have seen a slow-down in growth, which was close to zero for the 6-year period 1999–2005. However, there are large interannual variations, which, as yet, cannot be fully explained. The highly increased levels of atmospheric methane are due to continuing anthropogenic emissions, which are larger than natural emissions. Most important sources are wetlands, rice agriculture, ruminant animals, biomass burning and refuse dumps, but also industrial activities including all major fossil energy sources, coal, oil and natural gas.

In 2005 the atmospheric concentration of the 'laughing gas' nitrous oxide – N_2O – was about 320 ppb, almost 20 per cent above the pre-industrial value in 1750 (AR4, 2007). Concentrations have continued to grow more or less linearly during the last decades. Of singular greenhouse gases, nitrous oxide comes third in current global warming effects, but is lower on the list if we look at larger groups of gases, for instance, fluorocarbons. Increased concentrations are mainly due to human activities, particularly connected with agriculture and land use changes. The most important sources are the production and use of nitrogen-containing artificial fertilisers, but also the use of various fossil energy sources. Emissions are sharply increasing from the application of catalytic cleaning converters in cars, and may become important if private car use continues to rise globally. As shown in Table 2.1, nitrous oxide has a long lifetime, of about 120 years. In order for concentrations to be stabilised at current levels, anthropogenic emissions would need to be immediately reduced by more than 50 per cent. If emissions were held at current levels, concentrations could increase to about 400 ppb over several hundred years, thus substantially increasing their global warming effects.

Fluorocarbons are a group of strong greenhouse gases with a particular history in environmental policy. They are among those compounds used in fridges and freezers. Originally, the chlorine fluorocarbons (CFCs) were developed as alternatives to the former highly flammable and poisonous gases. They were considered to be ideal alternatives, without any adverse effects on health and local environment on the ground. However, in the early 1970s the theory of the serious roles of these gases in destroying the stratospheric ozone layer was launched (Nolin, 1995). These effects seemed to occur through leakages from the ground up into the upper atmosphere. After some years, the claims gained international support, and became the basis for the later Montreal Protocol, strongly restricting and abandoning these types of fluorocarbons. Alternative gases were developed by the industry, gases without the adverse effects on stratospheric ozone. Most important were the HFCs – the hydrofluorocarbons shown in Table 3.1. However, like the CFCs, they proved to be very strong greenhouse gases. When the leading environmental debate turns from stratospheric ozone to greenhouse gases and climate change, these gases should again come into critical focus. In a former work I have used this case to substantiate a thesis of the environmental problems of environmental solutions, which turn up when environmental problems are addressed and analysed within a framework of reductionism (Høyer and Selstad, 1993).

Many other examples can be given, which will be covered more exensively in later chapters. To reduce local air pollutions, catalytic converters have become obligatory for all new petrol cars. However, these converters cause very large increases in emissions of nitrous oxide, which is a very potent greenhouse gas. Restrictions on the application of the gas in hospitals have been enforced for mothers giving birth to children, but there are no similar restrictions in relation to the number of cars in use. Production of biofuels is another case. First of all they are produced to relieve us from fossil fuels and from CO_2 emissions in transportation. But that is not of much help when they can cause substantial increases in life-cycle emissions of other greenhouse gases, notably nitrous oxide generated from the wide use of artificial fertilisers. Analyses show that the total global warming potential of some biofuel alternatives may be larger than the fossil fuels they are supposed to replace.

Reuniting CO_2 with fossil energy

My claim is that efforts to disconnect CO_2 from fossil energy have become increasingly prominent in the current discourse. At first this may seem rather astonishing, more so as it applies to all the three major forms of fossil energy, coal, oil and natural gas. Not only is this the case for each separately, but also when the benefits of one are stressed in relation to the others. Looking at the 40-year history of the discourse on climate change and CO_2, such disconnection has largely turned up in the last 10–15 years. It was almost unheard of a couple of decades ago, when the terms sustainable development and sustainable energy were coined in the 1987 UN Brundtland Commission Report (WCED, 1987). In this context it is fair to emphasise that this report and its international follow-up processes gave a new, strongly reinforced impetus to the later highlighting of climate change issues and policies.

The disconnection takes many forms, technologically and politically. Claims are made that all three fossil sources can become CO_2 free or CO_2 neutral. Thus, for instance, coal is presented as the ultimate sustainable energy source of the future, as the global resources are so much larger than the other two. From a focus on the need to reduce, and terminate, coal production globally, the issue has turned to one of expansion, even as a condition for a sustainable future. In Norway, claims made by both the oil industry and the 'red–green' government are that policies to reduce CO_2 emissions have no bearing on Norwegian oil and gas production. They are pointed out as two completely separate policy arenas. Norwegian oil and gas are argued to be particularly low emitting, 'clean', and include the prospect of realising a future production system freed from all CO_2. The logic is that further expansion of just this oil and gas production represents one of the major contributions Norway can make to global combating of climate change. Along this line of argument, virtually any country with major fossil resources can come up with reasons for expanding *their* production of oil, gas or coal.

CCS – carbon capture and storage – is one of the new buzz words. Carbon sequestration is another term applied. Seemingly, this is a straightforward technological solution to the problems of CO_2; it is claimed that CO_2 can be captured, sequestrated and retained all together. However, not only does this require storage, but also storage of immensely large volumes. As with the long-term handling of nuclear waste, geological repositories are the most promising. They must be geologically stable and, besides the large storage volumes, posit a guarantee against smaller or larger leakages for centuries, even for thousands of years. Such guarantees are hard to give, especially when related to the extremely large volumes needed globally. Long-term repositories for nuclear waste, on the other hand, do not represent this type of volume problem. Their problem is primarily one of extremely high intensity in certain places.

Norway is a frontrunner in this approach to the climate change discourse. According to the 2007 New Year TV speech by the Norwegian prime minister, CCS is the 'moon landing' technology of our time; a moon landing where Norway can take a leading role (Stoltenberg, 2006). Two of the six new national research centres for 'sustainable energy' are actually in the field of CCS and the particular Norwegian prospects may seem promising. When their subsea oil and gas geological repositories are emptied, the repositories can, in principle, be filled with recycled CO_2 from onshore and offshore energy production plants. This also enables increased extraction of the subsea oil and gas resources, thereby contributing to an extension of CO_2 production. The new land to really conquer is the continuous future of Norwegian oil and gas export to the rest of Europe under the label of fossil energy guaranteed CO_2 free or at least CO_2 neutral.

As a main form of disconnection between CO_2 and fossil energy, CCS has several important limitations. The globally very limited storage capacity of satisfactory geological formations is one. Another is related to my former claim of disconnecting CO_2 from other climate gases. Both production and use of fossil energy are important sources of emissions of many other climate gases. CCS is only addressing CO_2 in isolation. And, with CCS, CO_2 is even disconnected from other environmental problems. Worldwide fossil energy production, transportation and use are well documented as major sources of environmental problems: nature conservation and land-use problems, air, sea and water pollution and health problems for humans. However, most basically, CCS represents the very classic end-of-pipe solution to environmental problems. This is where pollution problems seemingly are solved at the very end of the emission pipes. From all our modern environmental history, we know that this has proved very unsatisfactory, or even wrong.

During the last two decades, the need to develop source-orientated environmental solutions has gained great support, to the extent that we may talk about a discursive turn: solutions where problems are fundamentally addressed even before they are produced. This was expressed as a basic understanding in the Brundtland Commission Report (WCED, 1987), presented as one of the basic conditions for future sustainable development (Høyer, 1997, 1999). Another term coined in this same context was life cycle analysis (LCA), emphasising the need to address environmental problems from cradle to grave. CCS leans heavily towards the grave end. On the other hand, LCAs of complete coal energy chains, in particular, show how the larger CO_2 emissions, not to mention total climate gas emissions, are produced before the coal is burnt in the power plants, where they can be subject to CCS.

The disconnection and the CCS technology it presupposes are based on a point-source approach to environmental problems. Such point-sources are mostly large power plants, whether coal, oil or gas fired. Again, this is the classic way of addressing environmental problems. The two later decades have, however, highlighted the importance of diffuse sources. Such sources are all small, but there are so many of them that their sum total has become alarming. CO_2 emissions from the many diffuse sources cannot be captured, sequestrated and stored because of the many transportation means and their sum emissions. To a large extent, point sources and their particular problems belong to traditional modernity, the classic industrial society. Diffuse sources, on the other hand, express problems more dominating in late modern consumption societies. Thus there are reasons to claim that the very ideological foundation for the disconnection between CO_2 and fossil energy is out of tune with crucial societal development, which adds to the illusions of facticity in the solutions proposed.

Reuniting CO_2 with energy

Not only is CO_2 disconnected from fossil energy, my claim is that it has become disconnected from energy all together. Again, this takes many forms, technologically and politically. An elaboration requires a rephrasing: 'all energy issues are largely reduced to matters of CO_2 emissions'. With the limitation to fossil energy, this was substantiated under the previous thesis, entailing that all fossil energy issues are reduced to matters of CO_2 emissions. The disconnections are claimed to take place through both forms of reductionism. Some examples have already been given. Nuclear energy as an ultimate global energy solution is one. This is even marketed as sustainable nuclear energy, implying a reductionist understanding of sustainability as equal to CO_2 free. Such a position is also only possible when all environmental problems are reduced to matters of CO_2.

Development of renewable, alternative energy production is another case. We are faced with two quite different situations and energy chains: one for stationary and one for mobile use purposes. The stationary situations are, for example, industrial processes, and heating, ventilation, and electrical appliances in houses and other buildings. The mobile situations are all forms of transportation, whether of passengers or goods. And when applying the term 'use purposes', this is exactly what basic energy issues are all about. When developing alternative sources of energy production, the most crucial questions to ask are:

- Which types of purposes is the energy needed for?
- Which forms of energy are most efficiently able to serve these purposes?

- Which new sources of energy can most efficiently and environmentally benignly produce and transmit the required forms of energy?

Thus, electricity is only one among many forms of energy. It is also an energy carrier. Other energy carriers are chemical energy in the form of liquid and gaseous fuels and heat energy at various heat quality levels.

Within the paradigm of CO_2 reductionism, there is no room for such questions. They are even considered irrelevant. The only question asked is:

- How can we develop extensive CO_2-free energy production?

The main perspective is one of production, not efficiency, energy use or adaptations between production and user needs. Answers can then become utterly wrong. Nuclear energy can illustrate this: electricity is the only energy form and carrier. But there must be available: a high voltage grid, lower voltage distribution nets and a lot of low voltage users in order actually to put the electricity into application. It has taken the rich over-developed part of the world several decades to establish such an electrified infrastructure of production, distribution and use. The under-developed parts of the world, however, have quite another situation. Necessary infrastructure is absent. The number of users and actual needs for electricity are very limited and not pressing in relation to other energy forms. Besides the serious environmental problems it entails, building CO_2-free nuclear energy to solve development problems in poor countries may easily take the form of 'cutting butter with a power saw'. It is not exactly elegant, far from efficient, and more like the wrong answer to a wrongly posed question.

Now turning back to the alternative renewables again. Both the stationary and the mobile alternatives are currently subject to the one-sided production perspective. The overriding problems of volume are strongly under-communicated. Unrenewable, stored energy resources, fossil and nuclear energy, are characterised by their high concentration, their intensity. Large energy production can take place with only very small spatial requirements. Their problems of volume are related to the environmental effects of production, emissions of climate gases and to the generation of radioactive waste. Renewable energy resources, on the other hand, are highly dispersed and thus characterised by extensity. Large energy production can only take place with extensive spatial requirements. The actual spaces required can in each case be reduced through various technological means. Usually, however, this entails more severe environmental impacts and conflicts. There are limited production volumes available from renewable resources due to the limited spaces, or lands, which practically can be utilised.

The two concepts 'sources' and 'sinks' can help to explain these differences. Limits of unrenewable energy are, first of all, a matter of limits of sinks. The limit to emissions of CO_2 is a typical sink limit. For fossil energy they are much more pressing than the more distant source limits. In the case of fissile – nuclear – energy resources, real source limits are hardly present at all; there are only real sinks limits. Limits to renewable energy production, on the other hand, are first of all a matter of limits of sources. In their totality these resources may be more than large enough, but there are strict limits to their actual availability, as emphasised, caused by the limits of available land. In some cases, however, renewable energy may have pressing limits of sinks. As has been illustrated for biofuels in particular, this applies to mobile energy alternatives. There are strict land limitations to their production volumes, but they have also been shown to generate large emissions of climate gases, even larger than existing fossil fuels for some of the alternatives.

Internationally, in politics as in science, there is a broad agreement that global emissions of CO_2 need to be reduced by at least 80 per cent by 2050. This is considered necessary to keep the world within the limit of an increase in global mean temperature of not more than 2°C. It is no less than an illusion to believe that such extensive reductions can be made while at the same time keeping the current level of global energy consumption, or even accepting a continuing increase in this consumption. Problems of volume, and their related limits of CO_2-free sources, are so acute that conditions for upholding the volume of energy consumption are non-existent. The opposing illusion is based on a disconnection between CO_2 and energy, where they are addressed as totally separate matters.

It is worth emphasising that these claims are in line with the major conclusions and recommendations in the UN Brundtland Commission Report from 1987 (WCED, 1987). This was a key understanding, given their original term of sustainable energy, implying not only CO_2-free energy but substantially less energy production and consumption. The CO_2 issue was primarily understood as an energy issue. As background material to the main report, the Commission had carried out worldwide analyses of potential availabilities of renewable energy production. The conclusions were evident: their global potentials were far beneath the current level of energy consumption. A conclusion drawn was the immediate need to keep global energy production at that level. However, this was the level in 1980, 30 years ago, and the reference year of the Commission report. In order for poor countries to reach reasonable levels of development, it was thus recommended that the rich countries reduce their energy consumption by at least 50 per cent within 40–50 years. If complete global equity was to be achieved, reductions of at least 80 per cent would be necessary, all the time with 1980 as the year of reference. This understanding came to dominate the discourse some years into the 1990s. Then the whole issue became one of CO_2 and CO_2 reductions without any bearing on levels of energy production and consumption.

Reuniting CO_2 with consumption

In the wider context, CO_2 is disconnected from consumption in general and not just energy consumption. In the aftermath of the Brundtland Commission Report (WCED, 1987), terms like sustainable production, but especially sustainable consumption, turned up quite soon. Separate UN supported conferences were focusing on these terms (NME, 1994). This is hardly the case any more. The current discourse on CO_2 reductions is kept running as if this has no bearing on patterns or on levels of consumption. It has become a discourse of separation, and technological isolation, separated both from broader energy issues and from their more fundamental societal relations.

Historically, for almost a century, fossil energy consumption and ensuing CO_2 emissions have grown together with industrial production and consumption. Some may consider this to be largely coincidental, without any major connections between consumption growth and the actual energy sources applied or levels of energy consumption. However, there are solid arguments supporting claims that these are to a large extent a matter of systemic and permanent relations. In that case, it is hard to substantiate how at least an 80 per cent reduction in CO_2 emissions can be attained without any major implications for patterns or levels of consumption.

There is probably no large disagreement as to what extent fossil energy consumption historically has been an important driver in the development and growth of the industrial production society in its version of classic modernity. More open for discussion seems to be to what extent it plays a similar role in the growth of the consumption society of late modernity. Empirical proofs are, however, pressing. CO_2 emissions have continued to increase, even where fossil energy consumption for industrial production has decreased.

A main reason is found in increases of mobility. It is well known that personal mobility has continued to grow in the post-industrial consumption societies. This is the case for automobility, use of private cars, and not least in aeromobility, passenger transport by airplanes. Less known is that even goods mobility, including mobility of basic material resources, has continued to grow. The transport intensity of goods and commodities has been increasing nationally as well as internationally; similarly for the international and global flows of major material resources used in infrastructural development, buildings, and the industrial manufacture of commodities. An explanation seems to be that late modern consumption societies of course still consume industrial products and consumer products, while the production itself largely takes place elsewhere, in the less developed and under-developed parts of the world. Additionally, the volumes and spatial requirements of the consumption societies' infrastructures and buildings have never been larger.

'Post-industrial' thus is a term with two sides. When global chains, from cradle to grave, are included, the analyses show that these societies are just as material intensive as before. And never have their mobility intensities and CO_2 intensities been larger. Mobility, and the transportation activities it is based on, is *the* sector in all societies most heavily dependent on fossil energy. In other works, this author has elaborated on how the mobile society and the fossil society historically have grown together as Siamese twins. It is not possible to envisage how one, mobility, can continue to prosper and grow while disconnecting from the other, fossil energy.

Similarly, mobility and consumption are connected to each other with regard to levels as well as patterns of consumption. The late modern consumption societies are tied to mobility in several ways. Automobility is in itself a major form of consumption; however, it is also a precondition for other forms. Most late modern consumption cathedrals – shopping malls – are totally dependent on automobility. And aeromobility has increasingly taken the same form. Airports are huge cathedrals of consumption. And air travel for the sole purpose of consumption has become ever more frequent, in Europe as well as in the US. Leisure time consumption has gained particular importance, in volume, in the economy, and in all its conspicuousness. Consumption solely for leisure purposes has become a key driver in the current growth of consumption societies. Leisure consumption similarly is the form of consumption most heavily dependent on mobility, be it auto- or aeromobility.

Reuniting CO_2 with economic growth

The Limits to Growth was the title of a renowned 1972 book by a group of scientists at Massachusetts Institute of Technology, published under the auspices of the Club of Rome (Meadows *et al.*, 1972). Until then, the perspectives of most works within this field had been about *Problems of Growth*. The 1972 book highlighted an important discursive turn. Claims about the need to halt further economic growth and to develop a steady-state economy came to dominate the discourse for many years to come. The theoretical field in economics today known as ecological economy stems from this period.

In 1987 the UN Brundtland Commission did not seriously address these issues. Actually, it was one of the key topics in the report most evidently lacking a thorough analysis. And the Commission recommended further economic growth, even in the rich part of the world; however, it emphasised that a fundamentally new form of growth was required. Sustainable growth was the new term, which was to dominate the discourse in the years to come. According to the report, the world economy could continue to grow, while reducing energy consumption, climate gas emissions and other important environmental impacts.

Ecological modernisation has been the superior notion most commonly connected to such viewpoints, with Factor 4, Factor 10 and Factor 20 as more specific terms (Weizsäcker, 1998; NCM, 1999). Factor 4 entails a 75 per cent reduction in basic resource consumption and environmental impacts, while doubling the economy – doubling wealth – in rich countries. Factor 10, a 90 per cent reduction, is considered by many to be the long-term necessity. Factor 20 implies that this is combined with the doubling of the economy, thus in the long term requiring a 95 per cent (Factor 20) reduction in environmental loads from current levels.

The various factor proponents claim these load reductions to be both necessary and achievable. This seems unduly unrealistic. It takes quite some technological revolution, much more than a 'moon landing', to achieve 75–95 per cent reductions in resource consumption and environmental impacts. I have already outlined that global development after the Brundtland report shows no sign of such decoupling actually taking place. Impacts that were supposed to be reduced, even to a very large degree, have continued to increase largely in line with economic growth.

Indeed, a substantial critique of the sustainable growth term has been raised, very fundamentally by several world leading economists, among them Nobel Prize laureates. One of the key economists has on several occasions criticized sustainable growth as an impossibility theorem (Daly, 1989, 1993). A basic understanding expressed is that the world is overfull already and beyond the stage where it can take any increase in environmental loads. One thing is for certain: the world – the globe – is not becoming any larger. Most fundamentally, an economy in growth is something becoming larger. Something growing within the absolute limits of something not growing is impossible, especially when we are already beyond these limits. The need for 80 per cent or more reductions in global CO_2 emissions is only one of many indicators of a world already overloaded.

Exactly this last understanding perhaps was the most fundamental breakthrough of the Brundtland Commission report and its sustainable development concept. As outlined, limits to (future) growth was, until the report was published, the dominating way of understanding. However, when the MIT group of scientists published their follow-up book 20 years after their first one, it was entitled *Beyond the Limits* (Meadows *et al.*, 1992). This time their analyses emphasised that the world was far beyond the limit where there was still time to break down; the real matter was now about retreating, very soon and to a very large extent. An economy of retreat is something quite different from an economy of growth, and even from a steady-state economy. Such an economy is first and foremost one of *reduced volumes* in total.

For more than a century, fossil energy growth has been at the very heart of economic growth. This has been the case for oil as well as for coal and natural gas. All have had key roles in industrial development, while oil in particular has been a key driver in the further development of late modern consumption societies. Additionally, and due to the immense monopoly profits, money from fossil energy production has dominated investments to foster the worldwide economic growth machine. This fossil energy dominance in economic growth relates to some key physical characteristics of fossil natural resources. They are almost universal with their availability in many different parts of the world. They are, to a large extent, present as very concentrated rich resources. In many cases, they can be produced with only a little advanced technology, thus very different, for instance, from the production of nuclear energy from fissile resources. Concentrations of the resulting energy products are very high, with much energy per unit weight and volume. Most of the products, at least from coal and oil, can be readily stored and transported at ambient temperatures and pressures without any energy losses of importance. Products can easily be transported worldwide over very long distances, emphasising their universal and ready availability.

Expressing the close links between the mobile society and the fossil society, I have used the term Siamese twins. The close links between the economic growth society and the fossil society may be expressed through the same term. Economy cannot continue to grow alone. It must still grow together with fossil energy. Reductions in CO_2 emissions of at least 80 per cent are not achievable without similar reductions in fossil energy use *and* a new form of retreating economical development.

Reuniting CO_2 with sustainable development

My claim is that CO_2 has become disconnected from the wider concept of sustainable development. A rephrasing helps the further elaboration: sustainable development is largely reduced to a matter of CO_2 emissions. Some aspects of this have already been outlined. Anything deliberately made CO_2 free or CO_2 neutral is termed sustainable. On the same basis, nuclear energy is advanced as sustainable nuclear energy.

As they were developed by the Brundtland Commission in 1987, sustainable development and sustainability were both coherent and comprehensive terms. A rather problematic turn in this particular discourse has taken place over the last decade. Sustainable development (SD) is claimed to have a triple bottom line (TBL), consisting of ecological sustainability, economical sustainability and social sustainability more or less as separate spheres. Originally, this was presented by John Elkington, admittedly to make the SD concept more attractive to actors in business world. Thus the title of the work was 'Cannibals with forks – the triple bottom line of 21st century business' (Elkington, 1997). In former works, he applied terms like 'Win–win–win business strategies' and '3P – people, planet, profits' (Elkington, 1987, 1992).

To a large extent, TBL has become a new argument for business-as-usual in politics as well as in economics. The TBL concept carries an inherent notion of the three separate spheres, which can be balanced to each other. However, if the race of balancing is going to begin the starting point is not an ideal world of balance, as over a long period, the three have reached very different positions. Ecological but also social forms of sustainability have large handicaps in relation to the economic 'business-as-usual' form of sustainability ('profits').

Actually, the two superior concepts of sustainable development – the TBL and the Brundtland Commission version – are incompatible and not supplementary. On the one hand, there is the TBL concept with three quite separate spheres of sustainability, and where keeping a balance between the three is the major task. On the other hand, there is the original Brundtland concept where the integrative aspects of the three are highlighted, under the condition, however, that the task first of all is to secure long-term ecological sustainability.

The following question must be asked: what primarily characterises sustainable development, according to the original Brundtland Commission understanding? The answer provides an outline of what can be referred to as the major characteristics. They are found on three levels: extra prima, prima and secunda. These are terms borrowed from thermodynamics. Extra prima denotes energy (or other natural resources) at superior quality levels. Prima implies lower quality, but still very high. Extra prima and prima then are the main answers to the characterisation 'major'. Secunda has been included to put into context characteristics prevailing in the current debate on 'operationalisation' of the concept (Høyer, 1997, 1999).

Extra prima characteristics are:

- Ecological sustainability
- Satisfaction of basic needs.

Prima characteristics are:

- Nature's intrinsic value
- Long-term aspects
- Fair distribution of benefits and burdens globally
- Fair distribution of benefits and burdens over time
- Causal-oriented protection of the environment
- Public participation and democracy.

Secunda characteristics are:

- Reduction of today's total energy consumption in the rich countries
- Reduced emissions of climate gases, especially CO_2
- Reduction of today's consumption of non-renewable energy and material resources in the rich countries
- Increase of today's consumption of renewable energy and material resources
- Pollution levels within the tolerance levels of the ecosystems
- Giving priority to technological development for efficient exploitation of natural resources.

This final list could be very long (Høyer, 1997, 1999). The term secunda implies that they can be derived from extra prima and prima characteristics, and is what characterises reduced emissions of climate gases. This underlines the necessary conditions for reuniting CO_2 with sustainable development. The demand for reduced emissions of greenhouse gases is, first of all, necessarily based on a superior demand for ecological sustainability. However, any type of policy aimed at reduced emissions is not necessarily in line with sustainable development. In addition, fair distribution is needed, globally and over time, as well as being a priority in order to satisfy basic needs.

As to the extra prima characteristics of ecological sustainability, it is worth emphasising that the very term sustainability has its origin in ecological science. It was developed to express the conditions that must be present for ecosystems to sustain themselves in a long-term perspective. In the Brundtland Commission report, there are several references to the necessity of ecological sustainability, such as: 'The minimum requirement for sustainable development is that the natural systems which sustain life on earth, in the atmosphere, water, soil and all living things, are not endangered' and 'There is still time to save species and their ecosystems. This is an absolute precondition for a sustainable development. If we fail to do this, future generations will never forgive us.' This 'minimum requirement' means a requirement to sustain biological diversity, corresponding to the so-called diversity norm, which has prevailed in Norwegian eco-philosophy. The diversity of species, life forms and ecosystems must be sustained as a necessary, but not sufficient, precondition for sustainable development.

Satisfaction of basic needs represents the core of the development part of sustainable development. As with ecological sustainability, it constitutes a necessary precondition. The other characteristics have no meaning unless these two preconditions are fulfilled. This is the very basis for extra prima characteristics. Still, there is an important difference. Maintaining ecological sustainability is a negatively defining obligation. It is about restricting the extent of man-made encroachments in nature in order to maintain the necessary ecological sustainability. It is not a primary objective to develop maximum ecological sustainability at the expense of satisfying basic needs. As for the fundamental development part, it is, on the contrary, a question of a positively developing obligation (Næss, 1992). A large number of people do not get their basic needs satisfied today. These must be given priority, even if it may imply a reduction of the biological diversity. At the same time, the total population is too large and measures must be implemented to reduce the population if ecological sustainability is to be maintained in the long term.

The Brundtland Commission's report underlines the fact that a living standard beyond the necessary minimum to satisfy basic needs is only sustainable if all consumption standards, both present and future, are established in terms of what is sustainable in the long term. The majority of people in the rich world live far beyond the limit of ecological sustainability (WCED, 1987). A reduction in consumption levels is needed. Consequently, a ceiling has, in principle, already been put on the contribution these make in reducing ecological sustainability. Only a lowering of the ceiling is in line with sustainable development.

As for the core of the development part, the following may be emphasised:

- It presupposes measures for satisfying the basic needs in poor countries, as well as for reducing the consumption in rich countries.
- Further reductions in biological diversity are today only in accordance with sustainable development when it is linked to the satisfaction of basic needs.
- The latter point is also valid as a condition for encroachments on nature within, or carried out, by rich countries.

The prima characteristics 'Fair distribution of goods and burdens, globally and over time' relate to the basic needs. They, and the intermediate needs necessary to satisfy them, should be met worldwide by all future generations. In this context, the claim is that this should also be based on a principle of fair distribution, within each generation (intra-generational) globally and between all future generations (inter-generational). These are fundamental elements of a global ethics. This again relates to the issue of fair distribution of burdens. Environmental burdens are not equally distributed. The impacts are more serious in some parts of the world than in others, applying not least to climate change. Similarly, future generations are going to be hit harder than the present one. The implications are that the remedial actions, i.e. environmental policies, must be such that they take into consideration the situation for the least favored members of the global society, now and in the future.

Causal-orientated protection of the environment is placed among secunda characteristics. The Brundtland Commission report outlines two major approaches to environmental policy. The former is characterised as the standard programme, reflecting an attitude to environmental policy, acts and institutions with the main emphasis on environmental effects. The latter reflects an attitude focusing on the practice causing these effects: '*These two attitudes represent clearly diverging views both of the problems and of the institutions which are to deal with them*' (WCED, 1987). A distinction is made between effect-oriented and cause-oriented environmental policy. The Commission emphasises that it is the former that has prevailed until now, whereas it is the latter which must be included in sustainable development. In connection with the climate change issue, causal-orientated solutions will be the ones implying significant and long-term reductions in the emissions of climate gases. Reduced consumption of energy and reduced consumption of fossil energy, in particular, must be part of these solutions. In comparison, CCS, carbon capture and storage, and forestation to increase the natural absorption of CO_2 will be typical end-of-pipe, or effect-oriented solutions.

Uniting CO_2 with a post-carbon society

As early as in 1863, J. Tyndall expressed the opinion that changes in the CO_2 concentration in the atmospheric air could influence the climate near the earth's surface (SMIC, 1971). The question was studied by the Swedish chemist S. Arrhenius among others with a publication in 1896 (Arrhenius, 1896). Several contributions were published during the 1950s and 1960s. In his 1966 book *Science and Survival*, the American biologist and environmental author Barry Commoner was among the early scientific observers warning about the seriousness of CO_2 emissions and climate change (Commoner, 1966). A notable scientist with important studies from this period was the Swedish meteorologist Bert Bolin (Bolin, 1970). From 1974 he was already playing a key role in the international efforts to synthesise international research on climate change. He became a highly respected and influential leader of the IPCC from its beginning and through the whole decade, from 1988 to 1997.

The IPCC was jointly established in 1988 by the World Meteorological Organisation (WMO) and the United Nations Environment Programme (UNEP). Over the past 20 years they have issued four independent assessments of the state of knowledge. Their first assessment report (FAR) was published in August 1990, the second in December 1995 (SAR), the third in January 2001 (TAR), and the fourth and last (AR4) in November 2007. It is no exaggeration to claim that the main conclusions have become more certain and that climate change problems have turned out to be more challenging during these 20 years of assessments. This is supported by later research findings issued after the last assessment report (AR4), claiming that the former worst case scenarios may be close to what we can expect in the future.

Climate change has been a key issue at all global summits on environment and development. It is not well known that this was also the case at the first global UN conference on the environment in Stockholm in 1972. A main report of the study of man's impact on climate, entitled 'Inadvertent Climate Modification', had been commissioned and was presented at the conference (SMIC, 1971). With the Brundtland Commission report in 1987 (WCED, 1987), a process was organised with international and global summits on sustainable development. Every tenth year there is a global summit. The first one was in Rio in 1992, the second in Johannesburg in 2002, and with intermediate international conferences every fifth year. Through all of this process, climate change has been a key issue, initiated in the Rio summit with the global climate convention as one of its main results. This convention established the necessary legal and political framework for the later Kyoto protocol, and the ensuing processes to enforce major tightening up of the protocol, as is now expressed in the Copenhagen process in 2009.

With all these early warnings, serious scientific endeavours, political processes and global summits, what has it really been about? What is it we have been struggling with all these years? My final claim is that it has all been about liberation from the iron cage of fossil energy, and to prepare ourselves for a society after the fossil age – a post-carbon society. Actually, this term has attracted many adherents in recent years. It has become the lead title of many research seminars and workshops, of political conferences, and in the US even a whole institute providing guidance on how to achieve such a society (EC, 2008)

In my six earlier theses I have summed up some major implications of moving towards a post-carbon society:

1 Even though carbon is the metaphor applied, it is also about combating other climate gases, and addressing all climate gases in combination.
2 It entails a development where our societies become liberated from all forms of fossil energy, whether coal, oil or natural gas.
3 It is a development towards a low-energy society, where the levels of total energy consumption in the rich part of the world are very much lower than today.
4 It is a development where consumption patterns must be seriously changed, and where consumption levels in the rich part of the world are substantially reduced – a low-consumption society.
5 It is a 'beyond growth' development, towards a society with reduced economic volumes – an economical development substantially retreating from its current overloading of nature and natural systems.
6 Not least it is a sustainable development in accordance with the real meaning of the term.

I have, in particular, highlighted the issue of mobility. Nowhere are the connections to fossil energy as strong as in the case of mobility; the mobile society and the fossil society have grown together like Siamese twins. This applies to all three major forms of mobilities: movement of persons, movement of commodities and movement of natural resources and materials. All three are deeply challenged by development towards a post-carbon society, not only as regards patterns but also mobility levels. Liberation from fossil energy entails, first of all, substantially reduced levels of mobility. The title of a 2009 book is *After the Car* (Dennis and Urry, 2009). I prefer the term 'beyond'. A post-carbon society is a society *Beyond the Car*, *Beyond the Plane* and *Beyond the Global Mobilities* of commodities and materials.

References

AR4 (2007). *IPCC 4th Assessment*, Contributions from Working Group III – *Mitigation*. Cambridge: Cambridge University Press.

Arrhenius, S. (1896). The influence of the carbonic acid in the air upon the temperature of the ground. *Philosophical Magazine*, **41**, 237–276.

Bolin, B. (1970). Carbon cycle. *Scientific American*, **223**(3), 124–132.

Commoner, B. (1966). *Science and Survival*. New York: The Viking Press Ltd.

Daly, H.E. and Cobb, J. B. (1989). *For the Common Good*. Boston: Beacon Press.

Daly, H.E. and Townsend, K.N. (eds.) (1993). *Valuing the Earth: Economics, Ecology, Ethics*. Cambridge: MIT, pp. 267–273.

Dennis, K. and Urry, J. (2009). *After the Car*. Cambridge: Polity Press.

EC (2008). *Towards a 'Post-Carbon Society*. European research on economic incentives and social behaviour. *Conference Proceedings*, Brussels, 24.10.2007. EUR 23172 EN. Brussels: European Commission.

Elkington, J. (1987). *The Green Capitalists: How Industry Can Make Money – and Protect the Environment*. London: Gollanz.

Elkington, J. (1997). *Cannibals with Forks: The Triple Bottom Line of 21st Century Business*. Oxford: Capstone.

Elkington, J., Hailes, J. and Knight, P. (eds.) (1992). *The Green Business Guide: How to Take Up – And Profit From – The Environmental Challenge*. London: Gollanz.

Høyer, K.G. (1997). Sustainable development. In Brune, D., Chapman, D., Gwynne, M.O. and Pacyna, J.M. (eds.). *The Global Environment: Science, Technology and Management*, Vol 2. Weinheim, Germany: VCH Publ.

Høyer, K.G. (1999). Sustainable mobility: the concept and its implications. PhD thesis. Roskilde, Denmark: Roskilde University.

Høyer, K.G. (2007). Ecophilosophy and the current environmental debate. [*Sosiologisk Årbok*] *Annals of Sociology*, Norway, no 3–4, 2007.

Høyer, K.G.and Selstad, T. (1993). [*Den besværlige økologien*] *The Troublesome Ecology*. NordRefo report. (1993). 1. Copenhagen, Denmark: NordRefo.

Høyer, K.G. and Holden, E. (2005). The ecological footprints of fuels. *Transportation Research Part D*, **10,** 395–403.

Høyer, K.G. and Holden, E. (2007). Alternative fuels and sustainable mobility: is the future road paved by biofuels, electricity or hydrogen? *International Journal of Alternative Propulsion*, **1** (4), 352–369.

Høyer, K.G. and Næss, P. (2008). Interdisciplinarity, ecology and scientific theory: the case of sustainable urban development. *Journal of Critical Realism*, 7 (2), 179–208.

NME (1994). *Sustainable Consumption*. Symposium report, 19–20 January, 1994. Oslo: Norwegian Ministry of Environment.

Meadows, D.H., Meadows, D.L., Randers, J. and Behrens III, W.W. (1972). *The Limits to Growth*. Washington, DC: Potomac Ass.

Meadows, D.H., Meadows, D.L. and Randers, J. (1991). *Beyond the Limits: Confronting Global Collapse, Envisioning a Sustainable Future*. Norwegian edition (1992). Oslo: Cappelen Publ.

Nolin, J. (1995). [*Ozonskiktet och Vetenskapen*] *The Ozone Layer and Science*. Stockholm, Sweden: Almqvist & Wiksell International.

Næss, A. (1991). [Hva er Bærekraftig utvikling?] What is sustainable development? In Album, G., Eckhoff, T., Godal, J., Hoyer, K.G., Næss, A., Kvaloy Setereng, S., Sorenson, H. and Vinj, A. [*Supermarked eller Felles Framtid?*] *Supermarket or Common Future?* Oslo: Cappelen.

NCM (Nordic Council of Ministers) (1999). *Factors 4 and 10 in the Nordic Countries – the Transport Sector – the Forest Section – the Building and Real Estates Sector – the Food Supply Chain*. Copenhagen: Nordic Council of Ministers.

NOU ([Norges Offentlige Utredninger] Norwegian Offical Publication). (2006). [*Et klimavennlig Norge*] A climate friendly Norway. NOU, 18. Oslo: Official Publications from Ministries.

Randers, J. (2006). [Klimaproblemet – Hva bør Norge gjøre?] The climate problem – what should Norway do? Lecture, Oslo 4 October 2006. Oslo: Polyteknisk Forening.

SMIC (1971). *Inadvertent Climate Modification*. Report of the Study of Man's Impact on Climate (SMIC). London: MIT Press.

Solomon, S. *et al.* (2007). *The Physical Science Basis*. Contribution to Working Group I to the Fourth Assessment Report of the Intergovernmental Panel on Climate Change. Cambridge: Cambridge University Press.

Stoltenberg, J. (2006). [*Rødgrønn Månelanding*] *Red–Green Moonlanding*. 21 October 2006. Oslo: Dagens. Næringsliv.

WCED (1987). [*Vår Felles Framtid*] *Our Common Future*. Brundtland Commission report. Oslo: Tiden Publ.

von Weizsäcker, E.U. (1998). *Factor Four: Doubling Wealth, Halving Resource Use – A Report to the Club of Rome*. London: Earthscan.

4 The dangerous climate of disciplinary tunnel vision

Petter Næss

Introduction

Developing responses to human-made climate changes poses enormous challenges to humanity. Present trends of increasing greenhouse gas emissions must be reversed and the emissions radically reduced within a few decades if 'dangerous climate change' (Department for Environment, Transport and Rural Affairs, 2006; Hansen *et al.*, 2007) is not to occur. To a great extent, this will require a departure from prevailing technical solutions, political priorities, economic incentives, cultural traditions and ethical judgments.

Basically, the climate change problematique is about the interactions between human societies and the natural environment. Climate change influences natural ecosystems in numerous ways that are to a great extent unpredictable. Ecosystem degradation has, in its turn, several adverse impacts on human life and human societies. In some regions, the conditions for human settlement are likely to radically deteriorate, and some densely populated regions will probably have to be abandoned due to rising sea levels. Existing spatial and social inequalities will most likely be aggravated, as the most dramatic impacts of global warming are predicted to occur in the poor countries in Asia, Africa and Latin America. A steeply rising number of 'climate refugees' is likely to migrate to countries where the environmental and economic impacts of climate change are relatively lower. This has caused the foreign policy chief of the European Union to declare climate change as a 'major security risk' (BBC News, 2008).

This chapter argues that theories and applications in energy and climate studies need to be strongly based on interdisciplinary integration. Unfortunately, such holistic approaches are rare in academic as well as in political discourse. Using the so-called DPSIR scheme, originally developed as a framework for environmental indicators, but not limited to this purpose, the chapter seeks to demonstrate the need for interdisciplinary integration in the analysis of driving forces (D), pressures on the environment (P), state of the environment (S), impacts of environmental changes on human and non-human life (I), and social responses to environmental change (R), with energy use and climate gas emissions as the topical focus.

The DPSIR scheme is based on the earlier PSR model developed by OECD and has been adopted by the European Environmental Agency as a framework for describing interactions between society and the environment all along the causal chain from driving forces, pressures, state, impacts and responses (EEA, 2008). Thus, the scheme addresses problem generation and problem manifestations as well as attempts to solve problems. My point here is that disciplinary tunnel vision and narrowing-down of scope occurs along all these phases. Although our purpose is not to develop or use indicators, we consider a scheme inspired by the DPSIR model that is relevant for the purpose of discussing the unfavorable ignorance of multi-causality and multiplicity of interrelated mechanisms occurring at different stages of the relationship between human activities and climate change.

Here, a few reservations may still be relevant. Reality is, of course, much less simple and clear-cut than the five-step DPSIR model might suggest. Within one specific stage, there may, for example, be a multiplicity of more detailed driving forces, pressures, states, impacts and possible responses. For example, in the context of global warming, the category of 'state' arguably includes increasing global temperature, changing demarcation of regional climatic zones, changed precipitation patterns, changed ocean currents, changing ecosystems (vegetation and wildlife zones), melting glaciers, rising sea level as well as more extreme weather events (and several other aspects of the state of the environment). But these various aspects are also linked to each other causally, and some of these links manifest in feedback systems. For example, rising global temperature has a causal influence on precipitation patterns as well as the melting of glaciers. Changes in precipitation combined with rising temperatures cause changes in the composition and geographical location of climatic zones supporting certain macro-scale ecosystems (vegetation and wildlife zones). Melting ice can, on the other hand, cause changes in ocean currents, which in its turn creates changes in the regional distribution of temperature and precipitation zones.

In addition to the cause–effect relationships within each category of the DPSIR model, it may be open to discussion whether a certain phenomenon belongs to one or the other stage of this scheme. For example, transportation may be a driving force causing greenhouse gas emissions, whereas the development of transportation itself may be seen as a result of several direct or indirect causes such as local and regional land use patterns, mobility-enhancing technologies, economic globalization and economic growth (Danish Road Directorate, 2000; Goetz and Graham, 2004; Christensen *et al.*, 2007) – and the dynamics of the capitalist economic system. These causal factors are themselves causally tied to each other, including social structural as well as agential powers. Moreover, feedback mechanisms exist, sometimes across the stages of the DPSIR scheme. For example, melting glaciers and reduced snow cover in arctic areas will reduce the albedo effect[1] and increase the absorption of solar heat, thus leading to an additional rise in temperature. Methane released from thawing tundra is another example. A third example is changes in vegetation zones as a result of changes in the regional distribution of temperature and precipitation zones, which may themselves influence temperatures and precipitation patterns (Intergovernmental Panel on Climate Change, 2007). Finally, it may often be difficult to decide whether a certain social activity (e.g. changing land use patterns in cities) should be defined as a driving force or a response. Usually, a certain policy (e.g. inner-city densification) is pursued for a multitude of reasons, and although such a policy may be seen as a measure to reduce car dependence and hence be categorized as 'response', it may, at the same time, be pursued for a number of reasons unrelated to the climate agenda and may even make up a driving force in relation to some other climate-related problems (e.g. the location of new buildings to areas vulnerable to flooding as sea levels rise and hurricanes become more extreme).

The DPSIR scheme will thus be used only as a crude framework to distinguish between mechanisms located at different main stages along the causal chain from driving forces through pressures, states, impacts to responding new practices. Table 4.1 below shows some selected examples of driving forces, pressures, states, impacts and responses along different stages of the DPSIR scheme.

Different story-lines on climate change

In May 2008, the Danish National Right-wing Member of Parliament Morten Messerschmidt raised the following Parliamentary Question to the Ministry of the Environment: 'Does the Minister of Climate and Energy acknowledge that recent research findings about influences of changes in the albedo of the surface of the planet Mars on the climate on Mars implies that things other than carbon dioxide exert influence on how the climate develops?'[2] By the way the question was formulated, the intention was clearly to imply that, if climate variations were caused by albedo changes, then the theory of human-made climate change must be false. The scientific debate on climate change is often framed as being about identifying the single correct theory among several competing and mutually exclusive theories. This way of framing the debate is quite common, in particular among the so-called 'climate skeptics' (i.e. debaters raising doubt about the existence of any human-made climate change). For example, referring to the theories of solar activity, the TV program 'The great global warming swindle' characterized the theories on which the United Nations appointed Intergovernmental Panel on Climate Change (IPCC) has based its recommendations as highly speculative, uncertain and a mere bluff. A widespread liberal media ideology dictates that different sides in a debate should be given equally broad publicity and coverage. As a result, in the media, especially in the USA, 'climate skepticism' has been grossly overrepresented compared to its position within academia (Boykoff and Boykoff, 2004), thus legitimizing inaction and a 'wait and see until we have more certain knowledge' attitude. The elevation of the minor Danish political scientist and statistician Bjørn Lomborg to the status of 'climate policy guru' is an example of this.

In the current climate debate, two competing, broad story-lines dominate: climate change as unwarranted worry, and climate change as a serious but malleable threat. Table 4.2 shows some key assumptions and foci of each of these story-lines along the different stages of the DPSIR scheme.

Perhaps needless to say, the 'unwarranted risk' story-line uses scientific knowledge in a highly selective way, more or less actively disregarding types of

Table 4.1 Some selected examples of driving forces, pressures, states, impacts and responses regarding global warming along different stages of the DPSIR scheme

Driving forces	*More fundamental:* Profit-driven economy, consumerist culture, economic growth, population growth, power relationships. *More immediate:* Transportation, energy requirement in buildings, manufacturing, food production, deforestation
Pressures	*Human-made:* Greenhouse gas emissions (CO_2, NO_2, methane etc), reduced albedo, methane release from tundra. In addition a few involuntary mitigation mechanisms, notably sulfur emissions. *Not human-made:* Solar activity
States	Increasing global temperature, changing demarcations of regional temperature and precipitation zones, melting glaciers, rising sea level, changes in the composition and geographical location of climatic zones supporting certain macro-scale ecosystems (vegetation and wildlife zones), more extreme hurricanes
Impacts	*On human life:* Reduced possibility for food production in areas of desertification, famine, lack of water, geopolitical conflicts, war. Spreading of tropical diseases to wider regions of the world. Reduced possibilities for certain types of human economic and non-economic activities. Increased need for cooling in buildings but reduced need for heating in cool regions. Increased possibilities for certain types of farming. Rupture of cultural traditions rooted in existing climatic, vegetation and wildlife
	On non-human life: Extinction of species (many do not manage to move at a speed matching the changing demarcation of climate zones), changes (usually reduction) in local/regional biodiversity
Responses	*Mitigation, addressing the immediate driving forces:* Technological (dematerialization) within the fields mentioned as driving forces; social (social organization, social norms) aiming partly to enforce technological change, partly to influence behavior; political/institutional: legislation, taxes, regulation; ethical/cultural. Limits on physical mobility
	Mitigation, addressing more fundamental driving forces: Reducing and eventually halting (or even reversing) economic growth; changing the economic system; profound cultural change
	Adaptation: Building dikes; avoid building houses in flood-prone areas; changing to new agricultural products and other types of diets; migration combined with policies to integrate climate refugees (but also possible: increased hostility and security measures in order to prevent climate refugees from entering less non-affected areas)

knowledge that might imply that a departure from business as usual would be necessary. Metatheoretically, cognitive relativism lends support to this story-line. But also the 'serious but malleable threat' story-line has a number of 'blind spots' in the sense that certain types of relevant knowledge are disregarded or downplayed. This is apparent in the policy discourse (in the media, politics and

Table 4.2 Key assumptions and foci of the story-lines regarding global warming of 'unwarranted worry' and 'serious but malleable threat' along the different stages of the DPSIR scheme

	Unwarranted worry	*Serious but malleable threat*
Driving forces	Predominantly natural and fluctuating	Human activities (transportation, buildings, agriculture, manufacturing) dominated by carbon-intensive technologies
Pressures	Variation in the sun's magnetic field and solar wind are the main causes of climate change	Greenhouse gas emissions (notably CO_2), feedback pressures due to glacier melting and methane release from tundra
States	Some rise in temperature within a certain time period, but not likely to continue	Serious risk of dangerous and irreversible temperature increase
Impacts	May be positive (increased crops in cold regions) just as much as negative. Measures to reduce greenhouse gas emissions will be much most costly than the impacts of global warming	Predominantly negative. Focus on impacts for human life: famine, lack of water, flooding, hurricanes
Responses	Generally not much need for responses, apart from some adaptation to the changes that have already manifested themselves	Main focus on mitigation: more efficient technologies, increased use of non-fossil energy. But increasingly also adaptation: dikes, avoid flood-prone building sites, etc.

public administration) and not least in the inputs of various lobby groups. But blind spots exist also in the academic discourse. Partly, this reflects funding structures making it different to direct research activities toward certain issues. It also reflects the narrowing-down of the scope within particular disciplines.

In the following, I will provide a number of concrete examples of how current academic research and policy discourses at different DPSIR stages are largely dominated by mono-causal (or sometimes 'few-causal') approaches failing to take into account the full range of relevant mechanisms involved in the process of climate change. One could perhaps call this a sort of reductionism. Strictly speaking, reductionism is to assume that while an entity X apparently exists, it is nothing but Ys or parts of Ys (Hartwig, 2007a). What we are dealing with in this chapter is 'reductionism' in a much wider sense: the tendency to explain phenomena solely by means of theories and concepts belonging to one single discipline. Instead of reductionism, we shall use the term of 'disciplinary tunnel vision' (or sometimes simply 'tunnel vision') about the tendency of omitting causal mechanisms operating within strata or domains of reality other than the one(s) covered by one's own discipline.

Admittedly, some of the types of tunnel vision dealt with in the following may be just as much politically imposed as emanating from disciplinary borders, as, for example, when mainstream economists fail to consider the contradiction between climate responsibility and unlimited economic growth. However, such inherently ideological positions often penetrate the world views of dominant parts of a discipline, as in the case of mainstream (neoclassic) economics, or in various technological sciences. Other types of disciplinary tunnel vision may not at the outset be politically induced (e.g. the 'nature-blindness of traditional sociology, see Dunlap and Catton, 1983 and Benton, 2001), but may function as a legitimizing of business as usual policies in the context of environmental policy and climate change.

In critical realist terms, such monodisciplinary approaches fail to incorporate the different strata of reality, the multiplicity of causal mechanisms and the interdependency between geographical scales involved in the investigated phenomena. Often, such disciplinary tunnel vision tends to result in a relocation of environmental problems (temporally, spatially and topically) instead of providing an integrated solution. It also often obscures underlying structural causes of non-sustainable practices.

I have chosen to focus on six different examples, two of which address driving forces (one being more fundamental and the other more immediate). One example addresses impacts, and three address different response strategies. My examples here do not include pressures and states.

In addition to the six cases, some other examples of disciplinary tunnel vision will be discussed more briefly.

It should be noted that all the examples below refer to research and discourses considering human-made global warming as a real phenomenon. Five of the six examples could be placed under the story line of 'serious but malleable threat', whereas one example (cost–benefit analyses of global warming) is less clearly tied to this, as some of the studies employing this approach conclude that any negative impacts of a given rise in the global mean temperature are smaller than the benefits of inaction on greenhouse gas emissions. In fact, some of those who have carried out or cited such studies in support of their arguments, appear to be agnostic or even skeptical as to the existence of human-made global warming, and in this respect their cost–benefit analyses might thus be considered a 'Plan B' strategy in case the climate skeptics fail to convince the public and decision-makers that greenhouse gas emissions do not cause climate change. Some of the performers and proponents of cost–benefit analyses of global warming (but not all) could thus be said to belong to the story-line of 'unwarranted worry'.

Example 1: Economic growth (driving force, fundamental)

In the discourse under the story-line of 'serious but malleable threat', human activities within various sectors of society, such as housing, transport, manufacturing or agriculture, are seen as driving forces of greenhouse gas emissions. Almost invariantly, the analyses direct their focus toward the rising levels of emissions, and how much of these emissions could be considered 'excess' because they stem from processes where climate inefficient (i.e. energy-intensive and/or based on fossil fuels) technologies are used. To a lesser extent, there is a focus on the growth in the production and consumption of commodities and services. This growth is a function of both the production and consumption per capita and the size of the population.

The global level of emissions E can be expressed by the equation:

$$E = P \times A \times T$$

where:

P = the size of the global population

A = the average affluence level (measured in consumption or production per capita)

T = the efficiency of the technologies in use, measured in terms of greenhouse gas emission intensity (cf. Commoner, 1972; Holdren and Ehrlich, 1974).

Of these factors, the dominant discourse seems to focus quite one-sidedly on the *T* factor. Comparatively less attention is given to population growth, although a focus on family planning programs and the role of religious communities[3] as an obstacle to such initiatives are part of the discourse. The growth in the A factor (i.e. the average affluence level) is almost never questioned. And, if it is, there is seldom an analysis of the driving forces behind this growth, let alone the consequences for the economic system if growth were to be ended. Thus, the driving forces of greenhouse gas emissions are depicted predominantly in terms of inappropriate technologies. This 'technology tunnel vision' draws the attention away from the need to address also the other two factors of the $P \times A \times T$ equation. Especially, the climate-destructive role of economic growth is downplayed. Needless to say, this serves to direct social responses (stage R of the DPSIR model) toward technical fixes rather than social changes.

This disregarding of growth (notably economic per capita growth, but also population growth) as a climate culprit exists in the political debate, in the media, and to a high extent also in the academic debate on climate issues. Arguably, the narrow foci of separate disciplines (economy, technical disciplines, demographics and social sciences) that do not bring in perspectives from other disciplinary fields facilitates the concentration on only one of the *PAT* factors at a time. In the next step, the political discourse and the media then concentrate their focus on the disciplines dealing with the causal factors considered politically most relevant. This filtering is, of course, heavily influenced by power: the knowledge that is perceived as compatible with the interests of powerful groups is highlighted, whereas knowledge perceived as inconvenient or threatening to these groups (e.g. because it might raise doubt about the environmental sustainability of continual economic growth) is marginalized. Fairclough (2006, p. 58) characterizes economic growth as 'an *assumed good*' and questioning economic growth as a value as 'a scandalous thing to do in most contexts in most countries'. Such suppression would, of course, be possible even if knowledge production and communication were not separated into disjointed disciplines. It could, however, arguably have been more difficult for media and politicians to filter out parts of such interdisciplinary research, in particular if the conclusions of the research underlined how the level of greenhouse gas emissions depends on the combined effect of *T*, *P* and *A* factors.

The tunnel vision involved when economic growth is disregarded as a driving force of greenhouse gas emissions could be termed 'volume-blind tunnel vision', indicating that there is a one-sided focus on technological and other measures to reduce the emissions per produced or consumed unit, while the impacts of a steadily increasing volume of the economy on the amounts of emissions is ignored.

In addition to this widespread tunnel vision, a second form of tunnel vision is occurring frequently among the few debaters who do identify economic growth as a major driving force of global warming: blindness to the growth-dependence of the international capitalist economy. Most of the growth-critical participants of the climate change discourse discuss economic growth as though this was a thing you could opt for, or abstain from, within a capitalist economic system, depending on your value prioritizations. This belief disregards the growth dynamics inherent in a profit-driven economy based on competition and private ownership of the means of production (see, e.g., Kovel, 2002). This latter tunnel vision could be termed 'capitalism-blind tunnel vision'.

Example 2: The transportation sector (driving force, immediate)

Within the transportation sector, the discourse on greenhouse gas emissions is embedded in the broader discourse on sustainable mobility. According to Holden (2007), four main strategies to promote sustainable mobility can be identified:

- development of more efficient vehicles
- promotion of public transport
- encouragement of environmental attitudes and awareness
- and sustainable land-use planning.

Among these strategies, the main focus globally has been on vehicle technology, although measures within the three remaining categories have also been addressed to some extent, notably in some European countries. The growth in mobility itself is, however, seldom questioned. Instead, the discourse focuses mainly on how this growth could be made less carbon-intensive by shifting to low-emission vehicles and modes of travel with a low average energy use and greenhouse gas emission per capita. Again, we see a situation similar to the focus on *P*, *A* and *T* factors in the previous example. There is a strong focus on reducing the emissions per vehicle kilometer traveled (especially for cars) and some focus on trying to change the modal split towards public transport. However, there is much less focus on the fact that growth in the number of kilometers that persons and goods are transported has so far outweighed any benefits in vehicle efficiency and modal split by a solid margin (OECD, 2006; Tapio, 2005).

This lack of critical focus on the growth in mobility is not a phenomenon occurring only in the political domain and in the media. It also prevails within transportation research to a great extent. Again, research on vehicle technologies is carried out within disciplines and institutional settings separate from the more social science-based or planning-related disciplines that are investigating the possibilities for changing the modal split toward less carbon-intensive modes.

The tunnel vision involved in the dominant discourse on the role of the transportation sector as a contributor to greenhouse gas emissions could be termed 'technology tunnel vision', referring to the much stronger focus on the technological performance of the vehicle fleet than on other important parameters (like changes in the modal split between different modes of travel, or the growth in mobility).

Example 3: Cost–benefit analysis of global warming (impacts and their desirability)

In late 2006, the British economist Nicholas Stern published a report on the economics of climate change, prepared as a commission for the British government (Stern, 2006). He concluded that it would be necessary to spend 1 per cent of the Gross Global Product to reduce the emissions of greenhouse gases down to a level sufficiently low to avoid a global rise in temperature above 2°C . This cost would still, according to Stern, be far lower than the costs due to the environmental impacts of unmitigated global warming in a business as usual scenario. Stern has later (in 2008) revised his calculation, taking into account to the faster-rising temperature increase predicted in the latest report of the Intergovernmental Panel on Climate Change (2007), now holding that spending 2 per cent of the Gross Global Product annually would be necessary to avoid dangerous global warming (Jowit and Wintour, 2008). Even such an annual amount would be lower than the current annual growth in the global economy (and many of the spendings would probably also count as contributions to this growth, as they would involve production and services, e.g. within renewable energy technology). Stern's proposal would thus allow for continued economic growth, albeit at a somewhat lower pace than with business as usual (until the business as usual economy possibly collapses due to the impacts of dramatic global warming). Stern's analysis is based on the methodology of cost–benefit analysis but applies a considerably lower discount rate than is common in such analyses. Partly for that reason, his report has been subject to strong attack from more mainstream environmental economists, notably the renowned American economist William Nordhaus, who claims that the costs of trying to reduce greenhouse gas emissions are much higher than the benefits resulting from such policies. In a study published in 1994, Nordhaus has himself calculated that a 3°C rise in the global average temperature will result in only a 1.33 per cent loss in world output (Nordhaus, 1994). This calculation is based on a discount rate for long-term environmental impacts of 3 per cent, which is considerably lower than the 6–8 per cent discount rates common in a number of countries (including the USA, Great Britain and Denmark), but still higher than the discount rate used by Stern.

The discount rates used by Nordhaus as well as Stern are, however, chosen from a number of subjective and value-laden premises. The neoclassical critics of Stern's analysis are therefore right in pointing to the lack of any objective foundation for the chosen discount rate. However, this criticism turns back on the critics themselves to an even higher degree. The reasons for choosing a discount rate of 3 per cent annually, let alone discount rates of 6, 7 or 8 per cent, are far from objective. The justification for discounting future consequences is based on expectations of continuous economic growth: if we will be ten times richer in the future, paying a cost of 1000 dollars will impact our welfare much less than paying the same amount of money today. This justification thus depends upon the assumption that growth can go on continually without running into environmental limits.

- 3 per cent annual growth from now (i.e. 2008) until the end of this century will make the global economy fifteen times larger than it is today.
- 6 per cent annual growth will make it 213 times bigger than today.
- 8 per cent annual growth will result in a global economy 1188 times higher in 2100 than today.
- Even Stern's discount rate of 1.4 per cent annually presupposes a global economy in 2100 more than three and a half times bigger than in 2008.

Maybe it will be possible by energy efficiency measures and by shifting to non-fossil energy sources to reduce the global greenhouse emissions down to one-fifth or less of the present level (in line with recommendations by the UN Climate Panel) in spite of nearly a quadrupling of the production and consumption of commodities and services – this would require a reduction by factor of at least 18 in the greenhouse gas emissions per produced unit. But global warming is not the only negative environmental impact of economic growth.

- Will it be possible to reduce the encroachments on nature and the loss of soil for food production resulting from constructions of buildings and infrastructure to the same extent?
- Or the conversion of biodiversity-rich natural forests into agricultural areas or biofuel farms?
- Or the tapping of non-renewable raw materials?
- And is even a slow pace of continued conversion of finite natural areas into building sites or monoculture fields or forests defensible in a multi-generational perspective?

Disregarding the relationships between economic growth and other environmental impacts than global warming, even the 'green' environmental economist

Stern commits 'greenhouse gas tunnel vision' when applying cost–benefit analysis based on a positive discount rate for long-term climate change impacts.

The same of course also applies – and to a substantially higher extent – to Nordhaus and other mainstream environmental economists who do not only demonstrate 'greenhouse gas tunnel vision', but also appear to reduce the value of the natural environment to the stated willingness to pay (for a limited number of natural elements or resources) among selected business leaders and populations. This 'economic man fallacy' is in itself a kind of reductionism, as it reduces human nature to a totally self-interested and utility-calculating being (and also assumes an illusory level of information about the impacts of choices in the marketplace). Nordhaus' conclusion that three degrees increase in the global temperature would only lead to 1.33 per cent drop in the global product clearly indicates that the negative value attributed to the environmental impacts of this warming is very low. Such a conclusion reveals a fundamental ignorance about the functions of the various ecosystems as life support systems for human and non-human life, reflecting a discipline that has been characterized as being immune to any insight from the natural sciences (Fullbrook, 2004). Needless to say, such 'autistic economics' represents severe tunnel vision. Moreover, the neoclassical economics on which the cost–benefit analyses of climate change are based are extremely anthropocentric: nature has only a value to the extent that humans are willing to pay for it. The analyses are also blind to the distributions of burdens and benefits between different population groups, in spite of the fact that those who are likely to be hardest hit by global warming (e.g. poor people living in flood-prone areas like Bangladesh) have very low per capita emissions of greenhouse gases. Thus, tunnel vision is abundant in the discourse on the impacts of global warming: 'greenhouse gas tunnel vision', 'economic man tunnel vision', 'autistic economics tunnel vision', 'extreme anthropocentrism tunnel vision', and 'distribution-blind tunnel vision'.

Example 4: Biofuels (mitigation, cross-sector general approach)

Several technologies for supplying energy from non-fossil, renewable sources exist: solar energy, wind power, wave power and, not least, hydroelectricity. Among these, hydroelectricity has a long tradition as an energy source in countries like Norway, Sweden and Switzerland, and it is becoming increasingly important in several other countries where the amounts of precipitation and the topographic conditions provide a basis for utilizing the energy in descending water. The contribution from wind power is also growing, and in Denmark, for example, wind energy accounts for about 15 per cent of the electricity production. Over the last few years, biofuels have become increasingly often mentioned as a climate-friendly energy solution, especially as a replacement for gasoline and diesel in vehicles (either directly or through the production of hydrogen). At Scandinavian gas stations, gasoline composed of 15 per cent petrol and 85 per cent bio-ethanol is sold as an environmental alternative under a particular green label. The European

Union has stated that, by 2020, 12 per cent of all gasoline and diesel sold from gas stations within its member states should be based on biofuels (BBC News, 2007).

However, biofuels are highly problematic from a broader sustainability point of view. For one thing, the production of biomass necessary to cover a significant proportion of the world's energy use by biofuels would imply a strong pressure to convert natural forests and areas now used for food production into energy-crop areas. The quite modest production of biofuels carried out so far has already contributed to increasing food prices and hunger among poor people who can no longer afford food. Large-scale supply of biofuels would seriously conflict with concerns of food security and food affordability. It would clash with important environmental sustainability goals of nature conservation and protection of biodiversity. Moreover, replacing food-producing agriculture in the European countryside with energy crops may even aggravate global warming if Asian or African rainforests are replaced with farmland in order to compensate for the reduction in European food production.

When biofuels are launched uncritically as a solution to the greenhouse gas emissions problem of the transportation sector, we are again witnessing tunnel vision: technological solutions aiming to solve one particular problem are discussed isolated from their impacts on other parts of the natural or social environment. In this case, we could call it 'CO_2 tunnel vision', since other environmental impacts than carbon dioxide emissions (including nitrogen oxide emissions, loss of natural areas and biodiversity, and reduced food supply) are not taken into consideration. It also involves 'micro-scale tunnel vision', as the gain in terms of reduced CO_2 emissions from the individual vehicle or industrial plant may be counteracted and even outweighed by increased deforestation at a larger systemic scale.

Example 5: Vehicle technology (mitigation, within-sector general approach)

As mentioned in Example 2, the problem diagnosis regarding the greenhouse gas emissions from the transportation sector has focused predominantly on the energy efficiency of motors and the energy sources used for propulsion of vehicles, with comparatively less emphasis on increasing the proportion of transport carried out by climate-friendly modes of transport, and with almost no questioning of the growth in the amount of transport. Similarly, mitigation strategies within the transportation sector have – especially in the United States – concentrated first and foremost on vehicle technology. In Europe, there has also been some focus on increasing the share of travel carried out by public transport, but here too, a limitation (or even more radically: a reversal) of the currently rapid growth in the annual distance traveled per capita and the equally steep increase in goods transport has largely been excluded from the agenda.

However, even very optimistic levels of vehicle technology improvement will be insufficient to achieve sustainable mobility if transport volumes and the use of cars and aeroplanes continue to grow at present rates (Holden, 2007). The non-sustainability of such growth lies not only in its aggravation of greenhouse gas emission levels, but also in the impacts of alternative energy sources on natural areas and farmland (cf. the previous example), land consumption for transport infrastructure, and the need for scarce minerals and other material resources to produce the rolling (or flying or sailing) stock and the transport infrastructure (Høyer, 1999). Many of the vehicle technology solutions proposed (e.g. hydrogen fuel cells) are also quite inefficient in terms of greenhouse gas emissions mitigation, although no carbon dioxide is discharged from the vehicle itself (Holden, 2007). In a system perspective including the emissions connected with the production of hydrogen, biofuels or electricity for batteries, each of these technologies still imply considerable greenhouse gas emissions, with battery-electric motors being the most favorable and hydrogen motors the least favorable alternative. The hydrogen alternatives are in fact worse in a greenhouse gas perspective than the most energy-efficient types of the conventional gas-driven motors (Holden, 2007).

Moreover, a one-sided focus on vehicle technology disregards the many environmental criticisms made of the transportation sector long *before* climate change entered the agenda of planners and decision-makers: traffic accidents, barrier effects, noise, fragmentation of natural landscapes, deterioration of historical urban environments, facilitation of urban sprawl, and reduced accessibility to facilities for the parts of the population that are not able to drive a car.

The tunnel vision involved in the discourse on mitigation strategies within the transportation sector includes 'technical fixes tunnel vision' (downplaying the need for social and behavioral changes), 'micro-scale tunnel vision' (focusing only on the vehicle propulsion process and ignoring the greenhouse gas emissions in connection with the production of hydrogen or indirectly from biofuels production), and 'greenhouse gas tunnel vision' (ignoring the other negative social and environmental impacts of traffic growth).

Example 6: Urban planning (mitigation and adaptation, cross-sector local approach)

The development of the spatial structure of cities can influence the inhabitants' greenhouse gas emissions in several ways. Dense and concentrated cities require less motorized transport and depend to a lesser extent on the private car than do low-density, sprawling cities (Næss, 2006a). Building types associated with high density (apartment buildings) require, other things being equal, less energy for space heating and cooling than low-density building types (single-family homes) (Owens, 1986; Holden, 2007). And whereas improved public transport, better cycling facilities and improved conditions for pedestrians contribute to reduce the number of car travelers, increasing road capacity to make car traffic flow more easily contributes in the opposite direction (Mogridge, 1997; SACTRA 1999).

In the European discourse on sustainable urban development, the link between urban land use and travel has been recognized to a higher extent than among planners and decision-makers in the USA. In Europe, limiting carbon dioxide emissions from transport makes up an important part of the contemporary rationale for the 'compact city' ideal in urban planning. Although highly compatible with a transport-reducing and less car-dependent urban development, land-use principles contributing to lower energy use in buildings (notably reducing the construction of single-family homes) has been addressed to a much lower degree. Moreover, whereas considerable attention (and in many cities also funding) has been directed toward public transport improvement, most cities have at the same time increased their road capacity in order to make provision for expected growth in car traffic.

In almost no setting – political, professional or academic – has the desirability or necessity of continuous growth in the building stock been questioned. However, rising floor area per capita makes it increasingly difficult to avoid spatial urban expansion, which will increase the distances between urban facilities and make the inhabitants more dependent on motorized travel, especially by car. Growth in floor area per capita also implies that a larger building stock must be heated, lighted and cooled. Growth in the building stock thus contributes, other things equal, to increase greenhouse gas emissions and will 'eat up' a considerable part of the gains from low-emission strategies like energy-efficient building technologies or transport-reducing location of residences, workplaces and service facilities.

The discourse among urban planning practitioners and academics on climate change challenges has until recently focused almost only on mitigation strategies. Only during the latest few years have strategies for adapting to the climate changes inevitably facing us as a result of previous and current emissions (there is a time lag between emissions and the resulting increase in the global temperature) entered the urban planning agenda. Some of the mitigation strategies pursued (e.g. urban densification in old harbor or industrial waterfront areas) may actually conflict with important adaptation principles (to avoid flood-prone building sites) – at least if the densification is not accompanied with measures to protect the new buildings from floods (e.g. high dikes, or raising the new buildings on pillars). So far, such awareness has been almost entirely absent in the urban planning discourse on climate challenges.

Thus, within the field of urban planning, we can discern at least the following prevailing forms of tunnel vision in policies and strategies to meet the challenges of global warming:

- 'transport tunnel vision' (ignoring the influence of densification vs. sprawl on energy use in buildings),
- 'public transport tunnel vision' (ignoring the counterproductive effect of road capacity increases, seen in a greenhouse gas perspective),
- 'growth policies tunnel vision' (ignoring the contribution of building stock increases to increased greenhouse as emissions, and failing to explore strategies for meeting societal needs with lower or no growth in floor area per capita), and

- 'mitigation tunnel vision' (failing to address the need for adaptation to unavoidable climate changes, and ignoring possible conflicts between mitigation strategies followed and appropriate adaptation measures).

As mentioned above, transport-reducing urban developmental strategies have been accepted and adopted to a lower extent in the USA than in Europe. This also applies to mitigation strategies in general. In USA, there appears (to the extent that climate change is at all recognized as a threat) to be at least as strong a focus on adaptation strategies as on mitigation. This also holds true for urban development.

Other examples

The magnitude of climate change is greatly influenced by energy policies and technologies. There are considerable unrealized potentials for energy-saving as well as for energy supply technologies that may reduce climate gas emissions. Some energy technologies which may immediately seem promising may, however, have unintended consequences undermining their potential to reduce greenhouse gas emissions (Holden and Høyer, 2005; Holden, 2007). Some technologies may also, while possessing a real potential for reducing CO_2 emissions, cause severe harms to biodiversity and food production (e.g. biofuels, cf. above). This illustrates the more general fact that technological solutions aiming to solve one particular problem are often discussed in isolation from their impacts on other parts of the natural or social environment. An outstanding example of this is the framing of nuclear power as a 'climate-friendly' technology. Nuclear power implies serious unsolved security and waste problems threatening to cause environmental degradations of a similar magnitude to global warming. The need to reduce the combustion of oil, coal and gas does not make the problems of nuclear power evaporate. The type of tunnel vision involved in the above examples could be termed 'single-problem tunnel vision', referring to the tendency of trying to solve one single (environmental) problem without taking into consideration the influences of the proposed solution to other impact categories.

Technological solutions aiming to counteract global warming are also often discussed without taking into consideration the infrastructural and institutional frameworks necessary for their implementation. Moreover, technological approaches to climate change often do not pay sufficient attention to the social conditions necessary for potentially climate-friendly technologies to really contribute to reducing climate gas emissions. For one thing, technologies must be 'domesticated' by the users, i.e. they must be understood and operated in a correct way. Experience from low-energy housing shows that this requirement is not always met. Secondly, climate-friendly technologies need to be widely implemented. However, due to the inertia of existing infrastructures and vested interests of the providers of climate-adverse solutions, the potentials for renewable, climate-friendly and energy-saving technologies remain unutilized to a high extent. The tunnel vision involved in the above examples could be termed 'prototype tunnel vision' referring to the frequent neglecting of the system conditions necessary for turning a prototype solution into a widely implemented technology.

Often, lifestyle changes are pointed out as a requirement for more sustainable patterns and levels of consumption as well as an adoption of climate-friendly technologies. However, individuals who want to change their lifestyle in a less energy-intensive and climate-adverse direction face many structural and cultural barriers. Low-consumption ideas are countered by advertising and a generally prevailing consumerist culture. Attempts by individuals to reduce their climate impacts, e.g. by traveling less, are countered by increasing demands for mobility on the job market and an increased compulsion for employees to make occupational flights as participants of globalized networks. Appeals to individual climate awareness are likely to be of limited effect as long as material, social and cultural compulsions and incentives continue to push in the opposite direction. Depicting individual awareness-raising and behavioral change as the solution to global warming without addressing the numerous structural conditions constraining and influencing the actions of individuals might be termed 'voluntaristic tunnel vision'.

Summary so far

Table 4.3 summarizes the various kinds of tunnel vision identified in the foregoing examples. As we have seen, tunnel vision is abundant in the analyses of driving forces, impacts, as well as responses relevant to global warming. The same applies to some of the debates on pressures on the environment. For example, monocausal explanations acknowledging only non-human-made contributions to temperature increases, or focusing only on increased carbon dioxide emissions as the human contribution to global warming while neglecting other greenhouse gases like methane and N_2O. Examples of tunnel vision are less obvious in the analyses of states (i.e. the extent to which the global climate is actually changing and will continue to do so if current trends of greenhouse gas emissions continue). Some 'climate skeptics' have attempted to raise doubt about the measurements of temperature increases so far (among others, by referring to the possible bias of the so-called urban heat island phenomenon) and/or questioning the causal models on which predictions about further temperature increases are made. It is still doubtful whether these objections (or the counter-arguments made by mainstream climatologists) could be classified as examples of disciplinary tunnel vision. Nevertheless, for the discourse of global warming as a whole, disciplinary tunnel vision is fragmenting the debate to a great extent.

This disciplinary tunnel vision implies that important causes, impacts and relevant policy responses are being obscured and excluded. In particular, this leads to ignorance of the interplay between causal mechanisms focused on by different theories, operating at different strata of reality and/or typically studied by different disciplines or sub-disciplines. Observed events in open systems are the results of the operation of a multitude of causal powers and conditions. Hence there is a great risk of misinterpreting the influence of one particular causal mechanism

Table 4.3 Different types of tunnel vision identified in debates on global warming

Example	*Stage in the DPSIR model*	*Types of tunnel vision involved*
Economic growth	Driving forces, fundamental	Volume-blind tunnel vision, capitalism-blind tunnel vision, nature-blind tunnel vision
The transportation sector	Driving forces, immediate	Technology tunnel vision
Cost–benefit analyses of global warming	Impacts	Greenhouse gas tunnel vision, economic man tunnel vision, autistic economics tunnel vision, extreme anthropocentrism tunnel vision, distribution-blind tunnel vision
Biofuels	Response	CO_2 tunnel vision, micro-scale tunnel vision
Vehicle technology	Response	Technical fixes tunnel vision, micro-scale tunnel vision, greenhouse gas tunnel vision
Urban planning	Response	Transport tunnel vision, public transport tunnel vision, growth policies tunnel vision, mitigation tunnel vision
Other examples	Response	Single-problem tunnel vision, prototype tunnel vision, voluntarism tunnel vision

focused on by a particular theory or discipline, and of failing take account of simultaneously operating causal mechanisms 'belonging to' other theories and disciplines.

Needless to say, such monodisciplinary narrowing of the scope hampers explanation and purposeful action and is detrimental and dangerous for human as well as non-human life. Instead, the different dimensions, aspects and scales mentioned in the preceding sections need to be integrated if climate-responsible policies and measures are to contribute to truly reduce global warming. Analyses failing to make such integration across dimensions and scales tend to result in a *relocation* of environmental problems (temporally, spatially and topically) instead of a solution of the problems (Høyer and Selstad, 1993).

Theories and applications in energy and climate studies therefore need to be strongly based on interdisciplinary integration. Here, interdisciplinary research is understood as 'constituted on the basis of the integration of a number of disciplines into a research cluster which provides, or purports to provide, a new framework or understanding' (Hartwig, 2007b, p. 259). This is distinct from mere multidisciplinary research, which draws on more than one discipline, without challenging disciplinary identities, in order to study an object that transcends disciplinary boundaries (ibid.). The latter approach is of course valuable and necessary, but not sufficient to obtain a knowledge base for a climate-responsible development.

Causes of disciplinary tunnel vision

Why, then, is the discourse on climate change to such a high extent characterized by disciplinary tunnel vision, in spite of the obvious call for interdisciplinarity? Below, I will discuss quite briefly some possible reasons why monodisciplinary and reductionist approaches are so prevalent in spite of their serious shortcomings. (For a more comprehensive account, see Høyer and Næss, 2008.)

Several barriers appear to be hampering integration of knowledge across disciplines. One such barrier is the power play and competition between different academic disciplines for status and funding resources, tending to create cultures where other disciplines and research traditions are treated in a condescending way or simply ignored. There are long traditions of disciplinary research, with systems of reward encouraging researchers to stay within their disciplinary fields rather than moving across. Sayer (2000, p. 7) characterizes disciplines as intrinsically parochial and imperialist, tending to illicit 'reductionism, blinkered interpretations, and misattributions of causality'. Another reason, possibly related to the former, may be a strong tendency to fragmented analyses within separate policy fields, where problems are analyzed in isolation from their social and environmental context, and often without considering impacts beyond the jurisdiction of the sector authority responsible for the analysis. Disciplinary tunnel vision may also be legitimated by certain postmodernist ideas, in particular the claim that the 'great narratives' are dead (Lyotard, 1984). Within such a general view, the focus is often limited to the particular and the partisan, with a pronounced skepticism to the possibility of finding any truths overarching the subjective beliefs among subgroups in a pluralist society. Practitioners of different disciplines also make up subgroups and this perspective decreases the motivation and theoretical prospects for avoiding disciplinary tunnel vision.

The objections against the assumptions and explanations held by state-of the art research on issues relevant to global warming do not only target their truth or falsity. They also concern how important the influences of human activities are on the global climate, how feasible it is to implement more climate-friendly strategies, and the extent to which such strategies will gain popular support. Issues of importance, feasibility and acceptance may be important explanatory factors illuminating why certain kinds of knowledge are embraced or dismissed in decision-making processes. Power obviously plays an important role here. The burden of proof is often asymmetrically distributed. The power of segments of society involved in activities causing greenhouse gas emissions will be bolstered if the policies pursued within these fields are widely perceived as beneficial and based on solid knowledge. It will therefore be in the interest of actors economically or ideologically tied to certain policies to ensure that the impacts of these policies are widely held to be in line with social objectives. Knowledge claims raising doubt about the compatibility of dominating practices within these areas with social goals (in this case about environmental sustainability) will thus not be welcomed by these actors (Barnes, 1988, quoted in Haugaard, 2003). Not only knowledge claims, but also (sub)disciplines may be ideologically tied to the interests of particular groups. Filtering of knowledge also includes the disregarding by 'hegemonic' disciplines of knowledge from disciplines conveying 'inconvenient truths'. Examples of this include the refusal of neoclassical economics to accept insights from ecology, geophysics and other natural sciences about environmental limits to growth in production and consumption, and from sociology about the difference between wants and needs.

Notably, the general climate of disciplinary self-containment is favorable for the hegemonic neoliberal discourse, as it legitimates the lack of incorporation of insights from other sciences into mainstream economics. Without the general acceptance for 'autistic' disciplines, neoclassical economists would hardly be able to go unchallenged and continue their implausible assertions about the possibility of combining eternal economic growth with environmental sustainability – an assumption now serving as a main bulwark against the development of more radical initiatives toward sustainability.

But there are also metatheoretical causes of the general lack of interdisciplinarity. Many of the most influential metatheoretical perspectives virtually prohibit, or at best strongly discourage, the inclusion of insights about certain parts of reality (e.g. things that cannot be directly sensed and measured, monetarized, or are not part of the discourse). According to Bhaskar and Danermark (2006), both empiricism/naïve realism, strong social constructionism, neo-Kantianism and hermeneutics include presuppositions which, if acted upon, would make research into multifaceted and multi-layered topics (like the climate change problematique) practically impossible. Below, a few illustrations of the shortcomings of these metatheories will be given, focusing mainly on empiricism/naïve realism and strong social constructionism, with references also to some more general positivist views often accompanying empiricism/naïve realism (for a more in-depth account, see especially Bhaskar and Danermark, ibid., but also Høyer and Næss, 2008).

According to a *positivist* view, social science research should emulate research within the natural sciences as much as possible. Knowledge based on research where the observations do not lend themselves to mathematical measurement and analysis will then typically be considered less valid and perhaps be dismissed as merely subjective opinions. Needless to say, this encourages disciplinary tunnel vision in numerous fields relevant to global warning, especially as regards social structural and cultural conditions affecting greenhouse gas emissions and the way global warming affects different population groups in different contexts. In terms of disciplinary tunnel vision, positivism/empiricism could be said to legitimize various kinds of technological, technical fixes and prototype tunnel visions typically occurring within the engineering disciplines.

Strong social constructionism would on the other hand typically limit the scope of enquiry to the cultural processes through which certain phenomena come to be perceived as environmental problems, and would ignore the underlying structural mechanisms creating these phenomena nor their impacts on the physical environment. At best, strong social constructionism would be agnostic to whether we can know anything at all about reality beyond discourse (Buch-Hansen and Nielsen, 2005). At worst, strong social constructionism may pave the way for the purely idealist view that there *is* no such reality (Macnaghten and Urry, 2001; Dingler, 2005). In terms of disciplinary tunnel vision, strong social constructionism could be said to legitimate the earlier mentioned 'nature-blindness' and 'spatial blindness' that have characterized the discipline of sociology since the days of Durkheim and Weber (Benton, 2001).

The above-mentioned types of tunnel vision do not necessarily imply that mechanisms operating at other domains or levels are totally disregarded. However, tunnel vision does imply that the types of explanations, impacts or measures 'belonging' to the disciplinary tradition in question tend to be *a priori* privileged, prioritized and emphasized at the cost of other mechanisms (Bhaskar and Danermark 2006, p. 8).

Table 4.4 summarizes a number of metatheoretical positions and other phenomena that may serve to legitimize and promote the various kinds of tunnel vision discussed in the examples. Oppressive power, understood as the negative characteristics of power such as domination, subjugation, exploitation and control that can be identified in given social structures is a partial explanation of tunnel vision within all the examples. In some of the cases, notably the tacit assumption of continual economic growth as a given fact, the unquestioned growth in transport/mobility, and the practice of cost-benefit analysis of global warming, power is probably the most important explanation.[4]

Failure to see climate change as necessarily taking place within a laminated system (see below) is another main contributor to disciplinary tunnel vision. In the example areas, this takes two forms: partly, it entails disregarding the multi-*effect* influences of activities in open systems, and partly it entails disregarding the multiplicity of *causes* influencing states and events in open systems. In both cases, lack of interdisciplinary integration is legitimized. There is a widespread failure to realize the multitude of mechanisms, types of contexts and characteristic effects involved in the climate change problematique, and the intermeshing (conjunctive multiplicity) or differential (disjunctive plurality) operation of causal mechanisms in phenomena such as global warming (Bhaskar and Danermark, ibid.; Bhaskar 1975/2008). Due to its obvious logical flaws, few of those who fail to realize the need for interdisciplinary integration are likely to defend such actualism explicitly. Rather, Humean actualism lingers in the traditional paradigms of individual disciplines due to inertia and lack of critical metatheoretical reflection, supporting the view that it is unnecessary, a waste of time and is detrimental to intra-disciplinary deepening to involve perspectives from other disciplines.

Actualism has another implication. In the context of the climate problematique it serves to draw the attention away from the role of the capitalist economic system, with its inherent growth compulsion, as a key driving force of

Table 4.4 Metatheoretical positions and phenomena legitimizing various tunnel visions

Example	*Types of tunnel vision involved*	*Meta-theoretical causes of tunnel vision*
Economic growth	Volume-blind tunnel vision, capitalism-blind tunnel vision, nature-blind tunnel vision	• Mainly $power_2$ for volume-blind and capitalism-blind tunnel vision, involving construction of compliance by the exclusion of known alternative possibilities and socialization such that the worldview of individuals precludes an understanding of real interests • Actualist fallacy: possibility and necessity is reduced to an actuality of states of affairs and will to power • Ontological anthropism, legitimizing a disregarding of natural and ecological limits to growth • Economic reductionism, leading to the view that nature is entirely substitutable by human-made capital
The transportation sector	Technology tunnel vision	• $Power_2$ as above, leading the attention away from mobility increase and private motorized travel as important contributors to greenhouse gas emissions • Failure to see climate change as a 'necessarily laminated system'. This leads to a disregarding of the multicausal influences on states and events in open systems (in this case transport emissions), thus legitimizing monodisciplinary and mono-causal approaches
Cost–benefit analyses of global warming	Greenhouse gas tunnel vision, economic man tunnel vision, autistic economics tunnel vision, extreme anthropocentrism tunnel vision, distribution-blind tunnel vision	• $Power_2$ as above, giving hegemony to an analytical method based on utterly weak assumptions • Economic reductionism, leading to the view that nature is entirely substitutable by human-made capital • Failure to see climate change as a 'necessarily laminated system'. This leads to a disregarding of the multi-effect influences of activities in open systems (in this case other environmental impacts of economic growth than greenhouse gas emissions), thus legitimizing discounting of future environmental impacts through an assumption of continual economic growth • Anthropism and atomism legitimizing autistic economics • Confusion of needs with wants and purchasing power, legitimizing distribution-blind utilitarianism

		• Ethical anthropocentrism legitimizing a disregarding of nature's intrinsic value • Individualist reductionism (upward conflation) legitimizing market demand as a measure of social value • Implicit strong cognitive triumphalism in the assumption of fully informed market agents
Biofuels	CO_2 tunnel vision, micro-scale tunnel vision	• Failure to see climate change as a 'necessarily laminated system'. This leads to a disregarding of the multi-effect influences of activities in open systems (in this case impacts of biofuel farming on ecosystems, biodiversity, food production and CO_2 emissions caused by deforestation, thus legitimizing monodisciplinary and mono-causal approaches • $Power_2$ as above, leading the attention away from the impacts on nature, food availability and food affordability resulting from biofuel farming
Vehicle technology	Technical fixes tunnel vision, micro-scale tunnel vision, greenhouse gas tunnel vision	• Failure to see climate change as a 'necessarily laminated system'. This leads to a disregarding of the multi-effect influences of activities in open systems (in this case other environmental impacts of car and truck transport than greenhouse gas emissions, and impacts operating at other locations than that of the vehicle engine), thus legitimizing growth in these modes of transport • $Power_2$ as above, leading the attention away from the higher greenhouse gas emissions from private motoring than from public transport
Urban planning	Transport tunnel vision, public transport tunnel vision, growth policies tunnel vision, mitigation tunnel vision	• Failure to see climate change as a 'necessarily laminated system'. This leads to a disregarding of the multicausal (size and composition of the building stock as well as urban transportation) influences on states and events in open systems (in this case greenhouse gas emissions), thus legitimizing monodisciplinary and mono-causal approaches. It also leads to a disregarding of the multi-effect influences of activities in open systems (among others, increased risk of flood as a result of harbor front development), legitimizing densification as a context-independent solution • $Power_2$ as above, leading the attention away from growth in the building stock as a contributor to greenhouse gas emissions

Table 4.4 Continued

Example	*Types of tunnel vision involved*	*Meta-theoretical causes of tunnel vision*
Other examples	Single-problem tunnel vision, prototype tunnel vision, voluntarism tunnel vision	• Failure to see climate change as a 'necessarily laminated system'. This leads to a disregarding of the multi-effect influences of activities in open systems (in this case other environmental impacts of energy technologies than greenhouse gas emissions), legitimizing environmentally and socially highly problematic and dangerous energy solutions • Failure to see climate change as a 'necessarily laminated system' also leads to a disregarding of the multicausal (e.g. technological development, economic profitability, domesticization by users) influences on states and events in open systems (in this case climate-friendly energy technologies), thus legitimizing a purely technical approach to socio-technical issues • Atomism and 'upward conflation' legitimizing individualist voluntarism • $Power_2$ as above, leading the attention away from social constraints discouraging or rendering impossible climate-friendly individual behavior

climate change. Capitalism is socially constructed, as is the high priority given to economic growth by nearly all countries of the world. It is *possible* to construct a different society. Given the threat of irreversible and dangerous global warming, it is arguably also *necessary*. The actualist fallacy implies that possibility and necessity is reduced to an actuality of states of affairs and Nietzschean will to power (Hartwig, 2007c).

The anthropic fallacy is 'the exegesis of being in terms of human being' (Bhaskar, 1993/2008: 205) and entails the irrealist position that there is no natural world independent of our mind. This position feeds the fantasy that there are no ecological or environmental limits to human activities. In the context of this paper, anthropism serves to legitimize the 'nature-blindness' of neoclassical economics, in particular its failure to acknowledge any ecological limits to economic growth. Such 'nature-blindness' is also flourishing within sociology, in particular among those sociologists influenced by strong social constructionism.

The assumption of neoclassic economics (and willingness-to-pay investigations in cost–benefit analyses) that human agents are all fully informed market agents implies an extreme, individual-level degree of *cognitive triumphalism* (Hartwig, 2007d), entailing that reality is necessarily knowable to us. Failure to distinguish between on the one hand human *needs* and on the other hand *wants and market demand* (Assiter and Noonan, 2007; Dean, 2007) legitimizes a utilitarian focus only on the total amount of demand satisfaction. This disregards the question whether this actually leads to the satisfaction of real needs, as well as the distribution of need satisfaction between different groups and individuals in society. This occurs in particular in connection with cost–benefit analyses, but also serves to support the prevailing prioritization of economic growth rather than encouraging enquiry into distribution of affluence, and market-based prioritizations within particular policy fields, e.g. the transportation sector. Cost–benefit analyses – of global warming as well as within particular sectors such as transportation (cf. Næss, 2006b) – are also based on *individualist reductionism* (upward conflation) legitimizing the sum of individual market demands as a measure of social value. The emergent properties of society are implicitly denied, reducing society to the aggregate sum of its constituents and social value to the aggregate preferences of individuals who possess purchasing power enabling them to satisfy their demands in the marketplace. Atomistic individualism also legitimizes the voluntaristic appeal to individual lifestyle change (into more climate-friendly behavior) while disregarding current structural constraints and disincentives against such changes.

Finally, *ethical anthropocentrism* leads to a denial of nature's intrinsic or inherent value (A. Næss, 1993). Fueled by ontological anthropism, ethical anthropocentrism implies that nature (including animals, vegetation as well as ecosystems and landscapes) is denied the status of something that we as humans have responsibility for. Instead, nature is only seen as having a value in virtue of its utility for humans as raw materials (minerals, timber, food, etc.) or playground (tourism, outdoor recreation).

The contribution of critical realism

Compared with the above-mentioned reductionist positions, critical realism could play an important role as an underlaborer of interdisciplinarity and a bulwark against disciplinary *tunnel vision*. Critical realism, with its maximal inclusiveness, can allow causal powers at different levels of reality to be empirically investigated, and also accommodate insights of other meta-theoretical positions while avoiding their drawbacks. According to critical realism, concrete things or events in open systems must normally be explained 'in terms of a multiplicity of mechanisms, potentially of radically different kinds (and potentially demarcating the site of distinct disciplines) corresponding to different levels or aspects of reality' (Bhaskar and Danermark, 2006). The objects involved in explanations of climate change and the efficaciousness of possible response strategies belong partly to the natural sciences, partly to the social sciences, and are partly of a normative or ethical character. They also belong to different geographical or organizational scales. Events and processes influencing climate change must be understood in terms of both physical, biological, socio-economic, cultural and normative kinds of mechanisms, types of contexts and characteristic effects. Acknowledging that reality consists of different strata, that multiple causes are usually influencing events and situations in open systems, and that a pluralism of research methods is recommended as long as they take the ontological status of the research object into due consideration, critical realism appears to be particularly well suited as a metatheoretical platform for interdisciplinary, non-reductionist research. This particularly applies to research into climate change issues, where, as has been illustrated above, other metatheoretical positions tend to limit the scope of analysis in such a way that sub-optimal policies within a particular aspect of the climate change problematique are encouraged at the expense of policies addressing the challenges of global warming in a comprehensive way.

Bhaskar and Danermark (2006) contend that social life must be seen in the depiction of human nature as *four-planar social being*, which implies that every social event must be understood in terms of four dialectically interdependent planes:

a material transactions with nature
b social interaction between agents
c social structure proper
d the stratification of embodied personalities of agents.

Regarding issues of climate change, the categories a–d are highly interrelated; for example:

- The dependence of climate-relevant human actions on socio-economic and socio-cultural conditions
- the impacts of these actions on the natural environment

- the ramifications of climate and other environmental change on human health, material security and confidence in existing social structures and prioritizations
- and the possible resulting political and structural changes.

In situations of interrelatedness between the categories a–d, we are facing what Bhaskar and Danermark characterize as a 'laminated' system, where explanations involving mechanisms at several or all of these levels could be termed 'laminated explanations'. In such situations, monodisciplinary empirical studies taking into consideration only those factors of influence 'belonging' to the researcher's own discipline run a serious risk of misinterpreting these influences.

Moreover, according to critical realism, the different strata of reality and their related mechanisms (i.e. physical, biological, socio-economic, cultural and normative kinds of mechanisms) involved in climate change are situated in macroscopic (or overlying) and less macroscopic (or underlying) kinds of structures or mechanisms (Bhaskar and Danermark, 2006). For research into issues related to climate change, such scale-awareness is crucial. As has been shown above, many of the shortcomings of technological solutions to climate problems stem from failure to take into consideration impacts at other locations than the place where the technical device is being installed (e.g. tropical deforestation due to farmland expansion necessitated by reduced European food production as more cultivated land is utilized for biofuel production). Similarly, analyses of social aspects of climate change need to include both local and global effects, and combine an understanding of practices within particular groups with an analysis of how different measures and traits of development affect the distribution of benefits and burdens across groups.

A new and improved story-line

What would an alternative storyline look like, if the shortcomings of the dominating European storyline on climate change were to be corrected, incorporating insights from research integrating knowledge and perspectives across disciplinary borders? Table 4.5 provides a sketch of some additional themes. In general, the topics addressed under the storyline of 'serious but malleable threat' would be accommodated also within the new storyline based on interdisciplinary insights. But some new topics would also be added. In Table 4.5, only these new topics are mentioned explicitly.

A new story-line incorporating insights from interdisciplinary research would notably pay much more attention to the contribution of growth (in the economy as a whole, and in the consumption of commodities and services such as buildings, transport, meat, etc.), both as a driving force and as a target for social response. Not only would it be necessary to confront growth policies, but it would also be necessary to look critically at the underlying causes of the strong pressure towards ever-increasing consumption and production and the related promotion of a culture of consumerism. Most likely, the need for fundamental social change would enter the agenda as a result of such a critical analysis.

Table 4.5 Key assumptions and foci of a possible alternative story-line on global warming based on interdisciplinary understanding, compared to the story-line of 'serious but malleable threat'

	Serious but malleable threat	*Climate change as part of ecological crisis necessitating profound social change*
Driving forces	Human activities (transportation, buildings, agriculture, manufacturing) dominated by carbon-intensive technologies	As in 'serious but malleable threat' PLUS: • Economic growth as a main driving force, and capitalism as a system necessitating economic growth. Also focus on population growth • At a more immediate level: focus on mobility growth and growth in the building stock
Pressures	Greenhouse gas emissions (notably CO_2), feedback pressures due to glacier melting and methane release from tundra	As in 'serious but malleable threat'
States	Serious risk of dangerous and irreversible temperature increase	As in 'serious but malleable threat'
Impacts	Predominantly negative. Focus on impacts for human life: famine, lack of water, flooding, hurricanes	As in 'serious but malleable threat' PLUS: • Focus on distribution of impacts between population groups, and on consequences in terms of nature's intrinsic value • Rejection of cost–benefit analysis as a tool for assessing the desirability of policies to counter climate change
Responses	Main focus on mitigation: More efficient technologies, increased use of non-fossil energy. But increasingly also adaptation: dikes, avoid flood-prone building sites, etc.	As in 'serious but malleable threat' PLUS: • Consideration of a broader range of environmental and social impacts of technical response strategies • Focus on the need to set limits to growth in consumption, production and the economy in general • Focus on how response strategies influence the distribution of burdens and benefits, and the satisfaction of basic needs for everybody • Focus on the need for profound social change (structurally as well as culturally)

In the analysis of impacts of climate change as well as possible policy responses, reductionist economist approaches like cost–benefit analysis would be replaced. We would need broader analyses paying a high attention to principles of ethically fair distribution between population groups at a given time as well as between present and future generations. An interdisciplinary-based storyline would also be better suited to include concerns for nature as such, in contrast to the ethical anthropocentrism characterizing the present dominant discourse.

The alternative storyline sketched above would thus address transformation from unwanted, unneeded and/or oppressive sources of determination to wanted, needed and/or liberating ones (Bhaskar, 1986, quoted from Hartwig, 2007e), in contrast to dominant discourses. Increasing interdisciplinary understanding may result in more and more people interpreting climate change not as a single (albeit overwhelmingly serious) problem, but as a symptom of a profound ecological and civilization crisis. Overcoming tunnel vision can contribute to make the climate crisis a turning point, facilitating the awakening and awareness-raising necessary to create a climate for radical changes in the relationships between humans and nature as well as within human societies.

Pedagogy and interdisciplinarity

Some of the reasons for disciplinary tunnel vision and lack of interdisciplinary integration may also be traced back to the classrooms. From primary schools to universities, teaching and exercises have traditionally taken place within the curricula of separate disciplines. Problems are analyzed from the perspective of one particular discipline. To the extent that real-life examples are drawn on, the purpose is usually to illustrate mechanisms postulated by one specific theory, not to explain the multiplicity of causal influences that make an event happen or result in a particular situation.

Arguably, this way of teaching and learning may socialize pupils and students into a monodisciplinary, mono-causal way of thinking. The lectures in different subjects placed neatly after each other in the time-table may in fact make up a school in disciplinary tunnel vision.

Problem-based learning (pbl) in project groups might be a remedy against disciplinary self-sufficiency. If students work on real-life problems instead of problems formulated through the lenses of one particular discipline, they will usually be forced to employ perspectives from several theories and disciplines. Reality is considerably more multifaceted than the theoretical lenses through which a single discipline – or tradition within a discipline – views the world. Working with real-life problems and in contact with external 'problem-owners', the theories and explanations offered would continuously be confronted with contextual conditions and with the views from people representing other theoretical backgrounds, worldviews and interests. One could imagine that such contact with 'the real life out there' would function as a kind of 'vaccine' against digging oneself too deep down in monodisciplinary wells from which only a narrow slice of reality is visible.

There will, of course, anyhow be a need to explain different types of theoretical propositions in depth, and this probably requires concentrated attention to be given to the problem in question, without too much 'noise' from other aspects of reality. I therefore do not reject classroom teaching in different subjects as one way of communicating knowledge. However, this model needs to be combined with other teaching models more conducive to a combination and integration of theories from different disciplines. Here, I think problem-oriented project group work makes up an important supplement, especially when informed by a metatheoretical perspective facilitating interdisciplinary thinking. As outlined above, critical realism seems like a particularly fruitful platform for such interdisciplinary integration.

Note

1 The albedo effect is the tendency of light-colored surfaces to reflect more radiation from the Sun than dark-coloured surfaces do (Danmarks Nationalleksikon, 1994, Bind 1, p. 215).
2 Question to the Danish Folketing No. S-1763, 2007–2008, referring to an answer dated 22 May 2008 from the Minister of Climate and Energy to Question to the Danish Folketing No. S-1526, 2007–2008.
3 For example, in 1997, the Vatican published a document stating that the Catholic Church has always taught the intrinsic evil of contraception (Pontifical Council For The Family, 1997).
4 Bhaskar denotes such oppressive power 'Power$_2$', as distinct from 'Power$_1$', which refers to the general causal powers of human agency whose characteristics entail the possibility of human emancipation (Morgan, 2007).

References

Assiter, A. and Noonan, J. (2007). Human needs: a realist perspective. *Journal of Critical Realism*, **6**(2), 173–198.

Barnes, B. (1988). *The Nature of Power*. Cambridge: Polity Press.

BBC News (2007). *EU ministers agree biofuel target*. Web article February 15. http://news.bbc.co.uk/2/hi/europe/6365985.stm.

BBC News (2008). *EU set to agree emission cut plan*. Web article March 14. http://news.bbc.co.uk/2/hi/europe/7293436.stm.

Benton, T. (2001). Why are sociologists naturephobes? In Lopez, J. and Potter, G. (eds.) *After Postmodernism. An Introduction to Critical Realism*, pp. 132–145. London/New York: The Athlone Press.

Bhaskar, R. (1975/2008). *A Realist Theory of Science*. London and New York: Routledge.

Bhaskar, R. (1993/2008). *Dialectic: The Pulse of Freedom*. London and New York: Routledge.

Bhaskar, R. and Danermark, B. (2006). Metatheory, interdisciplinarity and disability research – a critical realist perspective. *Scandinavian Journal of Disability Research*, **8**, 278–297.

Boykoff, J. and Boykoff, M. (2004). Journalistic balance as global warming bias. Creating controversy where science finds consensus. *Fairness and Accuracy In Reporting (FAIR)*, November/December. http://www.fair.org/index.php?page=1978.

Buch-Hansen, H. and Nielsen, P. (2005). *Kritisk Realisme*. Roskilde: Samfundslitteratur.

Commoner, B. (1972). The environmental cost of economic growth. In *Population. Resources and the Environment*, pp. 339–363. Washington, DC: Government Printing Office.

Christensen, T. H., Godskesen, M., Gram-Hansen, K., Quitzau, M.-B. and Røpke, I. (2007). Greening the Danes? Experience with consumption and environment policies. *Journal of Consumer Policy*, **30**, 91–116.

Danish Road Directorate (2000). [*Trafikvækstens anatomi. Kvalitativ analyse af determinanter for persontransportvækst.*] *The anatomy of traffic growth. Qualitative analysis of determinants of growth in passenger transport*. Copenhagen: Danish Road Directorate.

Danmarks Nationalleksikon (1994). [*Den Store Danske Encyclopædi.*] *The Great Danish Encyclopedia*. Copenhagen: Danmarks Nationalleksikon.

Dean, K. (2007). Needs. In Hartwig, M. (ed.) *Dictionary of Critical Realism*, pp. 323–324. London/New York: Routledge.

Department for Environment, Transport and Rural Affairs (2006). *Avoiding Dangerous Climate Change*. London: Department for Environment, Transport and Rural Affairs.

Dingler, J. (2005). The discursive nature of nature: towards a postmodern concept of nature. *Journal of Environmental Policy and Planning*, **7**, 209–225.

Dunlap, R. E. and Catton, W. R. Jr. (1983). What environmental sociologists have in common (whether concerned with 'built' or 'natural' environments). *Sociological Inquiry*, **53**,113–135.

EEA (European Environmental Agency) (2008). *DPSIR*. In Environmental Terminology Discovery Service. Web glossary. http://glossary.eea.europa.eu/EEAGlossary/D/DPSIR, accessed July 18, 2008.

Fairclough, N. (2006). *Language and Globalization*. London and New York: Routledge.

Fullbrook, E. (2004). Broadband versus narrowband economics. Introductory chapter in Fullbrook, E. (ed.). *A Guide to What's Wrong With Economics*. London: Anthem Press.

Goetz, A. R. and Graham, B. (2004). Air transport globalization, liberalization and sustainability: post-2001 policy dynamics in the United States and Europe. *Journal of Transport Geography*, **12**, 265–276.

Hansen, J. *et al*. (2007). Dangerous human-made interference with climate: a GISS modelE study. *Atmospheric Chemistry and Physics*, **7**, 2287–2312.

Hartwig, M. (2007a). Reductionism. In Hartwig, M. (ed.). *Dictionary of Critical Realism*, pp. 405–406. London and New York: Routledge.

Hartwig, M. (2007b). Interdisciplinarity, etc. In Hartwig, M. (ed.). *Dictionary of Critical Realism*, pp. 259–260. London and New York: Routledge.

Hartwig, M. (2007c). Actualism. In Hartwig, M. (ed.). *Dictionary of Critical Realism*, pp. 14–16. London and New York: Routledge.

Hartwig, M. (2007d). Cognitivism or cognitive triumphalism. In Hartwig, M. (ed.). *Dictionary of Critical Realism*, p. 68. London and New York: Routledge.

Hartwig, M. (2007e). Emancipation. In Hartwig, M. (ed.). *Dictionary of Critical Realism*, p.157. London and New York: Routledge.

Haugaard, M. (2003). Reflecting on seven ways of creating power. *European Journal of Social Theory*, **6**, 87–113.

Holden, E. (2007). *Achieving Sustainable Mobility. Everyday and Leisure-Time Travel in the EU*. Aldershot, UK/Burlington, USA: Ashgate.

Holden, E. and Høyer, K. G. (2005). The ecological footprints of fuels. *Transportation Research Part D*, **10**, 395–403.

Holdren, J. P. and Ehrlich, P. R. (1974). Human population and the global environment. *American Scientist*, **62**, 282–292.

Høyer, K. G. (1999). Sustainable mobility – the concept and its implications. Ph.D. thesis. Roskilde/Sogndal: Roskilde University and Western Norway Research Center.

Høyer, K. G. and Næss, P. (2008). Interdisciplinarity, ecology and scientific theory – the case of sustainable urban development. Forthcoming in *Journal of Critical Realism*, **7**(2), 2008.

Høyer, K. G. and Selstad, T. (1993). [*Den besværlige økologien*] *The troublesome ecology*. Copenhagen: Nordic Institute of Regional Policy Research.

Intergovernmental Panel on Climate Change (2007). *Climate Change 2007: Synthesis Report*. Contribution of Working Groups I, II and III to the Fourth Assessment Report of the Intergovernmental Panel on Climate Change [Core Writing Team, Pachauri, R.K. and Reisinger, A. (eds.)]. Geneva, Switzerland: IPPC.

Jowit, J. and Wintour, P. (2008). Cost of tackling global climate change has doubled, warns Stern. *The Guardian*, Thursday 26 June.

Kovel, J. (2002). *The Enemy of Nature: The End of Capitalism or the End of the World?* London: Zed Books/Fernwood Publishing.

Lyotard, J.-F. (1984). *The Postmodern Condition: A Report on Knowledge*. Minneapolis: University of Minnesota Press.

Macnaghten, P. and Urry, J. (2001). *Contested Natures*. London: Sage/TCS.

Mogridge, M. J. H. (1997). The self-defeating nature of urban road capacity policy. A review of theories, disputes and available evidence. *Transport Policy*, 4(1), 5–23.

Morgan, J. (2007). Power. In Hartwig, M. (ed.). *Dictionary of Critical Realism*, pp. 372–373. London and New York: Routledge.

Næss, A. (1993). The deep ecological movement: some philosophical aspects. In Armstrong, S. J. and Botzler, R. G. (eds.). *Environmental Ethics. Divergence and Convergence*, pp. 411–421. New York: McGraw-Hill, Inc.

Næss, P. (2006a). *Urban Structure Matters. Residential Location, Car Dependence and Travel Behaviour*. London and New York: Routledge.

Næss, P. (2006b). Cost–benefit analyses of transportation investments: neither critical nor realistic. *Journal of Critical Realism*, **5**(1), 32–60.

Nordhaus, W. D. (1994). *Managing the Global Commons: The Economics of Climate Change*. Cambridge, MA: MIT Press.

OECD (2006). *Decoupling the Environmental Impacts of Transport from Economic Growth*. Paris: OECD Environment Directorate.

Owens, S (1986). *Energy, Planning and Urban Form*. London: Pion Limited.

Pontifical Council for the Family (1997). *Vademecum for Confessors Concerning some Aspects of the Morality of Conjugal Life*. Vatican City: Pontifical Council for the Family.

SACTRA (1999). *Transport and the Economy*. London: Standing Advisory Committee on Trunk Road Appraisal.

Sayer, A. (2000). *Realism and Social Science*. London/Thousand Oaks/New Delhi: Sage Publications.

Stern, N. (ed.) (2006). *The Economics of Climate Change*. London: HM Treasury, Cambridge University Press.

Tapio, P. (2005). Towards a theory of decoupling: degrees of decoupling in the EU and the case of road traffic in Finland between 1970 and 2001. *Transport Policy*, **12**, 137–151.

5 Consumption – a missing dimension in climate policy

Carlo Aall and John Hille

Introduction

Climate change policies and research have traditionally focused more on how greenhouse gas (GHG) emissions relate to production than to consumption. An illustration of this focus is the conventional and most widely used accounting method for measuring GHG emissions given in the UN Framework Convention on Climate Change (UNFCCC). The method was developed, primarily by the industrialised nations, for the implementation of the Kyoto Protocol (Helm *et al.*, 2007). The UNFCCC method takes a geographical approach to emissions' responsibility. All emissions generated from production within a country's territory make up that country's total emissions. A small part of consumption-related emissions are also included, primarily those derived from energy use in residential housing and the private use of automobiles and motorcycles.

GHG inventories are the reference point for formulating policies on reducing GHG emissions. In view of the importance of GHG inventories in climate change policy-making, there has been surprisingly little debate on the implications of various system boundaries for GHG inventories (Peters and Hertwich, 2008). However, in the ongoing debates on post-Kyoto frameworks, the role of consumption is increasingly being examined (Munksgaard and Pedersen, 2001; Bastioni *et al.*, 2004; Peters and Hertwich, 2006, 2008). There are at least two reasons to consider a more consumption-focused climate policy. First, the Kyoto Protocol ultimately considers average emissions per capita in defining reduction needs. Consequently, there is a climate-justice element in the protocol, which implies the participation of each individual in contributing to reducing GHG emissions. This, in turn, can only be considered from the point of view of consumption, even on a national level. Rich developed countries such as Norway might otherwise see a constant decline in emissions – despite increasing consumption levels – because of increased imports of manufactured goods (Munksgaard and Pedersen, 2001; Bruvoll, 2006). Second, in countries or regions that have established emission trading schemes, such as the EU, only large emitters are considered in trading, leaving major parts of the emissions unaccounted for.

Peters and Hertwich (2008) argue that a stronger focus on consumption-related GHG inventories could help solve international allocation issues concerning international transportation, carbon capture and storage, and carbon leakage as well as reduce the importance of emission commitments for developing countries. Furthermore, according to Peters and Hertwich (2008), consumption-related GHG inventories may also reveal new options for emissions mitigation, encourage greater use of environmental comparative advantage, address economic competition issues, and encourage technology diffusion.

In 2006, an Official Norwegian Report assessed the potential of reducing Norwegian emissions by 60 per cent in 2050 as compared with 1990. However, the introduction contained an important reservation (NOU, 2006: 5): 'A radical shift in the Norwegian way of life in a more climate friendly direction could deliver major reductions in future GHG emissions. The commission has nevertheless chosen not to recommend such a strategy, because among other things we believe it would be politically impossible to put into effect'. In 2007, the Ministry of Children and Equity (in charge of consumption affairs) and the Norwegian Ministry of Environment commissioned a study to address the issues that were not covered in NOU (2006), namely, the challenging issue of how to reduce consumption-related GHG emissions. Our chapter is based on the results from this report (Hille *et al.*, 2008).

The international scientific debate on climate change and consumption

Several authors have discussed the differences between production- and consumption-focused climate policies (Munksgaard and Pedersen, 2001; Ahmad and Wyckoff, 2003; Bastianoni *et al.*, 2004; Lenzena *et al.*, 2006; Bruvoll, 2006; Helm *et al.*, 2007; Peters and Hertwich, 2006, 2008), and at least four conclusions can be drawn from this literature.

1. The main focus of both climate change research and policy-making has been on production, for both estimating GHG emission inventories and creating GHG mitigation strategies.
2. Over the past 10 years, there has been a growing support for the idea of including consumption in climate change research and policy-making. This support is mostly a result of concern about 'carbon leakage', which refers to the potential danger of shifting emissions from countries with binding caps on emissions (the so-called Annex I countries) to countries without such binding caps.
3. A new approach in climate change research and policy-making is linked to the development of methods for making personal consumption-related GHG inventories (so-called climate calculators). This approach includes a focus on soft policy instruments for changing consumption patterns, for example, carbon labelling and other means of providing information to facilitate climate-friendly consumption.
4. A climate-justice approach has recently emerged to compete with the existing production approach. The idea is that all people should equally share

> the burdens of reducing GHG emissions. This approach has lead to increased discussion of global per capita emission quotas and is usually accompanied with a policy approach that gives more emphasis to hard policy instruments, such as taxation and regulation.

The ways in which GHG inventories are defined is critical in forming GHG mitigation policies, and reviewing examples of national GHG inventories with a consumption focus may yield new and helpful insights. Comprehensive national consumer-focused studies have been conducted for at least three countries: the United Kingdom (Helm *et al.*, 2007), Sweden (Naturvardsverket, 2008), and Denmark (WWF, 2008). Each of these studies shows the potential for vast differences between production- and consumption-related GHG emissions inventories.

According to official UNFCCC data, GHG emissions in the UK have consistently declined since 1990, and the country has met its 2012 target of reducing emissions by 12.5 per cent as compared to 1990. An analysis by Helm *et al.* (2007) revealed, however, that the major reasons for the decrease in UK emissions since the 1970s are changes in the fuel mix for power generation (since the 1990s) and a reduction in energy-intensive manufacturing since the 1970s and 1980s. Applying a consumption-related GHG inventory approach, including emissions from international aviation, international shipping, and emissions embedded in imports, Helm *et al.* (2007) estimated that emissions are 72 per cent higher than the official UNFCCC figures and showed that there actually has been a 19 per cent *increase* in UK emissions since 1990. Helm *et al.* (2007: 26) conclude their study by stating:

> Using the production accounting basis, the UK has enjoyed a 2.1 per cent per annum downward underlying trend in emissions intensity over recent decades. However, the task ahead may be much more daunting. Instead of a 2.1 per cent per annum underlying trend of decarbonisation, the economy's appetite for greenhouse gases may have been growing. If, when more robust data is collected, this turns out to be the case; climate-change policy will have to deliver a much stronger correction to change the course of the greenhouse gas economy.

The Swedish Environmental Protection Agency's estimate of consumption-related emissions in Sweden in 2003 is 33 per cent higher than that of the official UNFCCC inventory (Naturvardsverket, 2008). The Swedish study stresses the importance of analysing GHG emissions from the perspective of both consumption- and production-related GHG emissions and states, 'The perspective of consumption provides a better picture of how our own patterns of consumption affect climate' (Naturvardsverket, 2008: 13).

The World Wide Fund for Nature (WWF) studied Danish consumption in 2001 and concluded that Denmark had caused global CO_2 emissions that were 20 per cent higher than the level in the official UNFCCC inventories (WWF, 2008). Furthermore, the study shows that imports represented nearly half of the total consumption-related emissions and that this emission category could increase by 40 per cent from 2001 to 2006. The WWF study makes the following three recommendations for climate policy changes in Denmark: (1) Denmark, like Sweden, should start making official GHG emission inventories that measure all Danish GHG emissions to supplement the existing accounts, which primarily include only production-related GHG emissions; (2) the Danish government should work with the Chinese government to create a strategy so that Denmark and China can create a more climate friendly developmental path for both countries; and (3) the Danish government should actively pressure the EU to remove its own trade barriers on environmentally friendly products.

A theoretical demarcation between production- and consumption-related climate policy

Consumption is commonly understood to be the final purchase of goods and services, and every other commercial activity is some form of production. This definition is used as the basis for producing national statistics on consumption and production, and to some extent, it is also reflected in the way national climate policies are developed and GHG emission inventories are produced.

The Intergovernmental Panel on Climate Change (IPCC, 1996: 5) states that GHG inventories should 'include greenhouse gas emissions and removals taking place within national (including administered) territories and offshore areas over which the country has jurisdiction'. Although this definition seems reasonable, Peters and Hertwich (2008) point out that closer analysis reveals two important weaknesses. The system boundary in the IPCC definition differs from that used in the system of national accounts, thus making it difficult to compare GHG inventories with economic quantities such as gross domestic product (GDP) and thereby creating problems in allocating emissions from international activities. Furthermore, the IPCC definition is based on country-level production, which can lead to problems when considering international trade and resource endowments.

To address both production- and consumption-related GHG emissions and account for all emissions, we therefore need to develop a typology that clearly distinguishes between the two types of emissions. Such a typology should take into account the following dimensions: ownership of the production facilities (domestic or foreign); location of production facilities (domestic or abroad); nationality of the consumer (domestic or foreign); nationality of products and services to be consumed (domestic or foreign); and location of consumption (domestic or abroad). Considering all of these dimensions, we can create 12 categories of emission (Table 5.1).

Categories 1 and 3 include emissions from all types of production facilities located inland, but they are differentiated by domestic and foreign ownership of the facilities. These two categories are the main components of traditional national GHG inventories. Category 2 includes emissions from international shipping and factories located abroad. Category 4 is not relevant for any kind of national GHG inventory because it includes global emissions abroad from

Table 5.1 Typology for defining system boundaries for national GHG emission inventories

			Location of production or consumption	
			Domestic	*Abroad*
Ownership of production facilities	National		1 Inland emissions from national production facilities	1 Emissions abroad from national production facilities and international shipping
	Foreign		2 Inland emissions from production facilities owned by foreigners	2 Emissions abroad from production facilities owned by foreigners
Nationality of consumer	Products and services (p&s) produced inland	National	3 Inland emissions from national consumption of p&s produced inland inland	3 Emissions abroad from national consumption of p&s produced
		Foreign	4 Inland emissions from consumption by foreigners of p&s produced inland	4 Emissions abroad from consumption by foreigners of p&s produced inland
	Products and services (p&s) produced abroad	National	5 Inland emissions from national consumption of p&s produced abroad	5 Emissions abroad from national consumption of p&s produced abroad
		Foreign	6 Inland emissions from consumption by foreigners of p&s produced abroad	6 Emissions abroad from consumption by foreigners of p&s produced abroad

foreign-owned production facilities. Categories 5 and 9 include emissions from domestic consumption of products and services (public as well as private) produced inland and abroad, respectively. Categories 7 and 11 are similar, except they represent emissions from foreign consumers visiting the country in question, predominantly foreign tourists. Categories 6 and 10 include emissions from the consumption of products and services abroad. Category 10 will normally represent the larger amount of emissions, predominantly as a result of consumption while on holidays abroad. Category 8 is emissions from exports, and category 12 (like category 4) is almost completely irrelevant in terms of a national consumption-related GHG inventory. However, it does include emissions from foreign tourists on their way to and from the country in question.

Using the emission typology illustrated in Table 5.1, we can distinguish between three main categories of emission inventories:

1 The national emission inventory as defined by the UNFCCC, which includes categories 1 and 3 as well as some of the sources in categories 5 and 7 (that is, emissions from energy use in residential housing and private use of automobiles and motorcycles).
2 A clear-cut national production emission inventory, which should include emission categories 1, 2 and 3.
3 A clear-cut national consumption emission inventory, which should include emission categories 5 to 12.

The boundaries of these three main categories of emission inventories may differ when used in practice. For example, a sector-specific inventory for the case of tourism might include emissions from both national and foreign tourists and all their inland activities (emission categories 5, 7, 9 and 11) as well as emissions from the transportation of foreign tourists to and from the country in question (categories 8 and 12) and emissions from natives spending their holiday abroad (categories 6 and 10).

Methods applied in the Norwegian study

We applied this typology to GHG emission inventories for Norway (Hille *et al.*, 2008) but used a simplified version because of restrictions in data availability and limited time resources.

To assess production-related GHG emissions, we used data from the National Accounts Matrix including Environmental Accounts (NAMEA). These are integrated environmental accounts produced by Statistics Norway for the purpose of comparing economic and environmental data by means of combining national accounts data and emission statistics at the industry level. Emissions related to private households are not included. NAMEA data use the same economic definition of Norway as the one used in the national accounts. Emissions from shipping and international air transport are therefore included, but those from other economic activities abroad are not (for example, from production facilities abroad owned by Norwegian economic interests). Emission category 2 is therefore only partially covered.

Calculating the consumption-related emissions was much more complicated, basically consisting of four main steps: (1) assess the amount of consumption in monetary or physical units from national statistics; (2) calculate the direct and indirect energy use caused by Norwegian consumption; (3) calculate by means of emission coefficients energy-related GHG emissions based on input gathered in step 2; and (4) calculate non-energy related emissions caused by Norwegian consumption. Total GHG emissions caused by Norwegian consumption are the sum of the amounts obtained in steps 3 and 4.

Consumption data were derived from three main sources of information: the biannual national survey of consumer expenditure carried out by Statistics Norway, the periodically conducted National Travel Survey carried out by the Institute of Transport Economics, and import and export statistics produced by Norwegian Customs and Statistics Norway.

In assessing the energy use, we began by estimating end use of individual energy carriers and then estimated the primary consumption of energy sources, using conversion factors obtained from previous studies combined with other statistical sources. Where end use of energy was in the form of electricity, we differentiated between electricity that was consumed directly in Norway and electricity used to produce imports of European or global origin. We assessed eight categories of consumption-related end use of energy: (1) direct use for personal transport; (2) indirect use for personal transport, including the energy cost of vehicles and infrastructure; (3) direct use in dwellings; (4) direct use for commercial and public services; (5) indirect use for these services (i.e. energy use for production of inputs); (6) energy use for construction and maintenance of buildings in Norway; (7) energy use for food consumed in Norway; and (8) energy use for all other goods (i.e. goods other than food, vehicles and energy).

Only categories 3, 7 and 8 are almost entirely and unequivocally related to Norwegian consumption. Some travel by Norwegians is related to export business, just as some business travel by foreigners is related to production chains that end up in Norwegian consumption. Energy use for the former should be deducted and that for the latter added to Norwegian indirect energy consumption, but we made the simplifying assumption that the two cancelled each other out. Some of the services produced in Norway are exported, either being consumed by foreign tourists or (as can be the case with consultancy and financial services, for example) delivered to companies abroad. Conversely, Norway also imports services. These activities affect emissions from categories 4 and 5. Here, we likewise made the assumption that net indirect imports of energy related to services were zero. This is not unreasonable in light of the fact that Norway's trade in services (excluding transport) is approximately balanced in monetary terms. In the case of buildings (category 6), we introduced a corresponding simplification: some buildings in Norway are used to produce goods for export and some buildings abroad are used to produce for export to Norway, but again the net effect was set to zero. The errors introduced through these simplifications are probably minor.

Almost all public services and the great majority of other services are produced for domestic consumption. Construction and maintenance of buildings turned out to represent about 5 per cent of consumption-related energy use, declining somewhat through the period we studied. The export/import factor mainly affects buildings for the manufacturing industry, which comprise less than 15 per cent of the building stock in Norway. Even a large relative 'trade deficit' in industrial buildings would be unlikely to substantially affect the overall results.

Using these simplifications, we could extract data for categories 3 and 4 (direct energy use in Norwegian dwellings and services) directly from the Norwegian Energy Accounts. In the case of category 1 (direct use for passenger transport), the extent of travel abroad by Norwegians (as opposed to travel within Norway) could be estimated and allocated by mode by combining statistical and other data, especially the previously mentioned regular surveys conducted over the entire period by the Norwegian Institute for Transport Economics. Data for energy use categorized by energy carrier per passenger kilometre for each mode of transportation were taken from previous studies, as were estimates of energy use per passenger kilometre for vehicles and infrastructure (category 2). In the case of food (category 8), estimates of overall energy use for the various production steps (production of capital goods and major inputs, farming and fisheries, food processing, marketing, and transportation between steps) were derived from a combination of statistical sources and previous studies. Energy use for domestic production for the domestic market (which is conveniently bounded in Norway because the country exports very little food and is almost 100 per cent self-sufficient in most animal products) was analysed separately from that for imports. A similar step-by-step procedure was followed in the case of buildings (category 6). The two remaining categories – inputs to services (5) and consumer goods excluding vehicles and food (7) – are dominated by imports. Figures for these were estimated on the basis of previous input–output analyses and international data on trends in energy intensities in manufacturing and transporting goods through the period studied, accounting for changes in the source of the imports. A complete list of sources is available in Hille *et al.* (2008).

Once we had estimated the end use of energy, split by energy carrier, the next step was to allocate these values to primary energy sources. If the carrier was a fossil fuel, add-ons from end use to primary energy were derived from Ecoinvent (2005). For biomass, end use was assumed to be equal to primary energy. The case of electricity is much more complex, not only because it can be generated from a wide variety of sources, but also because the generation efficiencies vary between countries and over time. We simplified the process by allocating electricity consumption to three regions: (1) Norway, (2) other OECD countries in Europe, and (3) the rest of the world. We then calculated source-by-source, primary-to-end-use efficiencies for each of these areas for the entire period. Add-ons for construction and maintenance of power plants and transmission lines were primarily taken from Ecoinvent (2005). We thus arrived at estimates for primary use of all fossil fuels, whether it was used directly or as an input to generate electricity. Inputs to generation of district heating were also calculated.

It is debatable whether there should be a distinction between consumption of electricity produced in Norway and that produced in the rest of OECD-Europe. Norway is formally part of a common Nordic market for electricity and, through a large number of transmission cables, part of an even wider European market. The argument for considering Norway as a separate unit is that Norway is normally a net exporter of electricity, so that all domestic consumption may be regarded as having been domestically generated. Which approach is used has major consequences on the outcome because 99 per cent of Norwegian electricity is generated by hydropower. Hydropower in Norway has traditionally been cheap, and because of this, energy consumption by Norwegian dwellings and services is overwhelmingly dominated by electricity. In Hille *et al.* (2008), we therefore present two alternative estimates of primary energy use: one based on the assumption that electricity consumed in Norway had been generated in Norway and the other (which was 53 per cent higher) based on the assumption that this electricity corresponded to a European mix. All results presented in this paper are based on the former assumption.

Once the primary energy use by fuel was calculated, the next step was to calculate energy-related GHG emissions. We did this by using fuel-specific factors for emissions of CO_2 equivalents (GWP 100) derived from SFT (2005).[1]

Finally, we estimated the non-energy-related GHG emissions caused by Norwegian consumption. Six categories of non-energy emissions were considered: (1) emissions of GHGs other than CO_2 and GHG precursors from aircraft; (2) process emissions of CO_2 in production of cement used for construction in Norway; (3) process emissions of CH_4 and N_2O from agricultural production for Norwegian consumption; (4) emissions of N_2O from production of fertilisers; (5) emissions of CH_4 from municipal landfills in Norway; and (6) other process emissions from companies that manufacture products for Norwegian consumption. It could be argued that the first category should be considered as energy related, but the distinction is not important for the purpose of our study. Agriculture was by far the most important source of non-energy emissions, followed by aviation. Process emissions from Norwegian agriculture were taken directly from national statistics; those relating to imported food were estimated via estimates of the amount of land used to produce the imports, and a specific add-on was used for rice imports. In the case of aviation, the additional GWP per passenger kilometre was taken to be 0.72 times that resulting from CO_2 emissions alone in 1987, rising to 0.8 times in 2006 as a result of a larger share of long-haul flights. Norwegian landfill emissions were also taken directly from published statistics, and process emissions for fertiliser and cement production were based on Norwegian data regarding specific emissions (Norway is a net exporter of both products). Process emissions from other manufacturing activities were derived from IPCC estimates of global emissions, using the simplified assumption that Norway's share in these emissions was equal to the country's share in global consumption.

A comparison of Norwegian production- and consumption-related GHG emissions

In this section we present the results of the GHG emission inventories for Norway for the period 1997–2006 (Hille *et al.*, 2008). The estimates for the three main categories of GHG emissions for 2006 are shown in Figure 5.1: Norwegian production (in Norway and abroad), Norwegian consumption (in Norway and abroad) and foreign consumption of Norwegian export products (limited to oil and gas exports). The UNFCCC inventories of national GHG emissions are also shown. National production emissions are dominated by emissions from Norwegian shipping (50 Mt CO_2e), which is not currently considered to be part of the standard national inventory. The remaining production-related emissions are dominated by the production of oil and gas (28 per cent); metallurgic, chemical, mineral, and oil refinery industries (23 per cent); and public transportation including aviation (15 per cent). National consumption emissions are dominated by domestic consumption of food (27 per cent), consumption of commodities other than food (19 per cent), domestic transportation (16 per cent), and air transportation abroad (15 per cent).

A number of insights can be gained from the results shown in Figure 5.1. First, unlike in Sweden, Denmark and the United Kingdom, consumption-related emissions in Norway are less than half those related to production. There two primary reasons for this: the size of the Norwegian shipping industry and the fact that Norway still has a large amount of economic activity within traditional GHG–emission-intensive industries (for example, metallurgic industries) in addition to a large oil- and gas-producing sector. Second, as Figure 5.1 clearly shows, the export of gas and oil is by far the largest contributor to 'Norwegian' GHG emissions.

The studies of Denmark and the UK previously noted that consumption-related emissions had increased whereas production-related emissions had decreased. As Table 5.2 shows, this is also true to a lesser extent in Norway. There was an annual average decline of 1.4 per cent in production emissions and an increase of 1.6 per cent in consumption-related emissions from 1997 to 2005/2006. The two categories of consumption that increased most are transportation abroad by plane and consumption of commodities other than food.

From 1985 to 2005, Norway experienced an explosion in the amount of air transportation. Whereas domestic transport by automobile (measured in person kilometres) increased by 41 per cent, domestic air transport increased almost five times that amount (by 189 per cent). However, the real 'explosion' occurred in air transport abroad on regular airlines (that is, excluding chartered flights), which increased by 863 per cent in the same period. Whereas air transport accounted for 3 per cent of the total amount of personal transportation in 1985, it accounted for 16 per cent in 2005, making it the second largest transport category after automobiles. Norway has also experienced large increases in the collective category 'commodities other than food'. For example, the money spent on clothing and shoes increased by 147 per cent and that spent on furniture and household equipment increased by 87 per cent from 1987 to 2005.

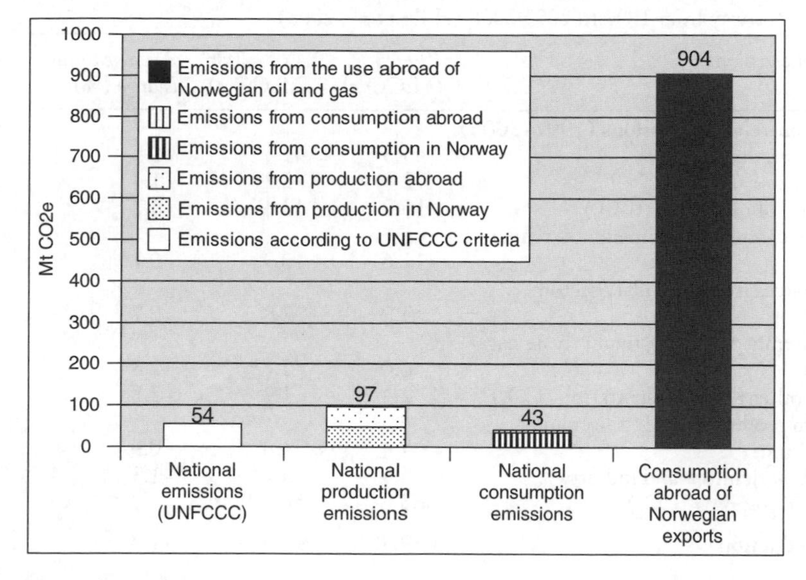

Figure 5.1 Different categories of GHG emissions for Norway in 2006 (Hille *et al.*, 2008).

Conclusions and policy recommendations

Inevitably, the level of GHG emissions will differ between a production-related emissions inventory (as defined by UNFCCC) and a consumption-related emission inventory. More importantly, the use of consumption-related GHG emission inventories may reveal new sources of GHG emissions and thus expand the GHG mitigation policy agenda. The previously mentioned Swedish study serves as an example. That study shows an 80/20 split between private and public consumption-related emissions and allows for a differentiation between four main activities and their related share of private consumption emissions: food (25 per cent), housing (30 per cent), travel (30 per cent), and shopping (15 per cent, with the purchase of clothes and shoes being the largest sub-category in this group). Thus, to reduce GHG emissions from private consumption effectively, the study points out the following changes that need to be promoted but that are currently not central to Swedish climate policy: change food habits, make housing more energy-efficient through the use of new technology and increased use of renewable energy resources, switch from private to public transportation, and increase the use of more energy efficient private cars for short- and medium-range journeys. The authors make the following interesting comment about long-haul journeys (Naturvardsverket, 2008: 13): 'there do not appear to be any technical solutions at present to limit the climate impact of aviation to a sufficient extent for extensive flying to be possible'.

Using the typology presented in Table 5.1 as a basis, we suggest a derived typology for the main areas of a consumption-related climate policy in high-consuming societies (Table 5.2). These policy areas are not generally covered by existing climate policies or national GHG inventories.

Table 5.2 Development of production- and consumption-related GHG emissions in Norway from 1997 to 2005/2006 (Hille *et al.*, 2008)

Sources of emissions	*1997 (MtCO_2e)*	*2005–2006 (Mt CO_2e)*	*Average annual change (%)*
Production-related emissions (1997–2005)			
Domestic production	45.3	47.3	+0.5
Primary sector (CO_2)	2.5	2.2	–1.4
Production of oil and gas (CO_2)	10.7	13.4	3.2
Metallurgic, chemical, mineral, oil refinery (CO_2)	11.6	10.9	–0.7
Other sources from secondary sectors (CO_2)	3.0	2.8	–0.7
Domestic transport other than private car (CO_2)	6.0	7.1	2.3
Services other than transportation (CO_2)	2.4	1.9	–2.8
N_2O from production of fertiliser and agriculture	4.4	4.8	0.9
CH_4 from agriculture and industry	4.7	4.2	–1.3
Norwegian shipping (CO_2)	63.8	50.0	–2.7
Total production	**109.1**	**97.3**	**–1.4**
Consumption-related emissions (1997–2006)			
Domestic transportation	6.6	6.9	0.5
Transportation abroad by airplane	3.9	6.3	6.8
Consumption of food	10.9	11.4	0.6
Housing	3.4	3.6	0.7
Commodities other than food	5.9	8.3	4.6
Private and public services other than transportation	5.6	5.3	–0.5
Emissions from municipal dumpsites	1.3	1.0	–3.0
Total consumption	**37.6**	**42.9**	**1.6**

Inland private consumption is on the climate policy agenda to a limited extent. The emissions from heating in residential housing and the private use of automobiles are part of existing national GHG inventories; hence, these policy areas are included in current climate policies. The coverage is not broad based, however. For example, a good deal of attention is focused on reducing GHG emissions from residential heating in Norway, but other climate-related aspects of housing have gained very little attention, for example, using low-emission building materials. Similarly, support for public transportation has received a great deal of attention, but measures directed at limiting the use of private cars have not. Measures related to food have similarly received little attention in climate policies, although the climate benefits of organic food and local produce have been discussed, and there is an emerging debate on the climate labelling of food (and other products).

Most high-consuming societies have experienced a strong increase in the import of commodities. Sub-categories of consumption, such as clothing, footwear and electronic items have experienced very large increases in the last decade. At the same time, however, these sub-categories have received very little attention in the climate policy debate except for the debate on climate efficiency branding of some electric and electronic items.

Public procurement is perhaps the consumption-related policy area that so far has gained the most attention in high-consuming societies. Several policy initiatives at different levels of government in many countries have addressed how to include environmental concerns in public procurement. Still, the focus in most cases has been limited to energy use (mostly for heating), the use of more efficient office equipment and, to a limited extent, transportation (e.g. promoting the use of video conferences as an option to long-distance travelling). Other aspects of consumption have received less attention.

Different strategies for promoting more sustainable forms of tourism have been part of the tourism debate for many years for both national and international tourism. However, these strategies have mainly focused on the negative effects of the 'stationary' parts of tourism activities (staying at hotels, dining at restaurants etc.), with the 'greening' of hotels achieving the most attention by far in Norway. The actual issue of the 'mobile' parts of tourism; that is transporting tourists – particularly the transportation to and from the destination – has received the least attention. Furthermore, the sustainability debate about tourism has primarily been focused at the industry and not the individual level. There seems to be great potential in developing strategies and measures directed towards changing tourism consumption, especially with regard to transportation.

Exports from one country will inevitably appear as inland consumption of imported goods and services in the recipient countries (categories 1 to 4 in Table 5.3), which is an argument against including this category in a consumption-related approach to climate policy-making. However, as argued in the WWF study of Denmark, the producer (the exporting country) and the consumer (the importing country) have a common responsibility for reducing GHG emissions relating to the products. Furthermore, it could be argued that, in the case of export-related emissions being very much higher than the 'remaining' national emissions (e.g. the case of gas and oil exports from Norway), there

Table 5.3 Suggestions for new main policy areas to supplement existing production-focused climate policy areas

<table>
<tr><td></td><td></td><td colspan="3">Location of consumption</td></tr>
<tr><td></td><td></td><td colspan="2">National</td><td>Abroad</td></tr>
<tr><td rowspan="3">Nationality of the consumer</td><td rowspan="2">National</td><td>(1) Inland private consumption</td><td>(2) Imported private consumption</td><td>(3) Public procurement</td></tr>
<tr><td colspan="2" rowspan="2">(4) National tourism</td><td>(5) Tourism abroad</td></tr>
<tr><td>Foreigner</td><td>(6) Exports</td></tr>
</table>

is an extra responsibility on the exporting country to reduce such emissions. This argument could be formulated as one of 'user responsibility'. This principle has been implemented in other contexts, for example, in the case of regulating weapons exports. One could imagine similar arrangements for oil-and-gas-exporting countries in the shape of agreements between the exporting and importing countries aimed at reducing GHG emissions from the oil and gas, for example, by implementing energy-saving programs.

The main reason for including consumption-related areas in climate policy is that an increasing share of the production of goods is taking part in developing countries while consumption is still increasing in the rich countries, resulting in a large increase in transportation of both goods and people. Interactions between production and consumption activities have become much more complex, making it more difficult to develop effective measures to reduce GHG emissions from this complex system. It has become very complicated even to measure GHG emissions; especially to align actual GHG emissions with specific production or consumption activities. It also has become harder to discern which emissions are 'hidden' behind the consumption of certain types of commodities.

The issue of consumption has long been controversial in the climate change debate, and this point is nicely illustrated by the famous statement made by former US President George H. W. Bush during the Earth Summit meeting in Rio de Janeiro in 1992: 'The American Way of Life is Not Negotiable. We cannot permit the extreme in the environmental movement to shut down the United States. We cannot shut down the lives of many Americans by going extreme on the environment'.[2] Bush's statement illustrates the long-standing controversial nature of applying restrictions on consumers to reduce GHG emissions.

Even so, the issue of consumption has slowly risen on the climate change agenda over the last decade in many high-consuming societies. In March 2008, the report that forms the basis for this chapter was presented at an open seminar arranged by the two ministries that had commissioned the report and with both ministers present. In a press release, the two ministers stated that, 'Taking care of the environment and climate must also have consequences regarding our consumption. Consumption has to a very little degree been part of a coherent environment discussion, and the means to be discussed have been directed towards production. Now the time has come to address consumption'.[3] Sadly, one year after this statement, there has been little substantial movement in Norwegian climate policy towards developing a specific consumption focus. In Norway at least, the debate on changing consumption to reduce GHG emissions remains primarily an academic one.

Notes

1 Chap. 4: 21; N_2O: 310.

2 Cited in *The Guardian* on 1 June 1992. The statement was made during a press conference, in response to a question about whether the United States would sign the Climate Convention (which it did not). Citation available at http://www.mail-archive.com/gep-ed@listserve1.allegheny.edu/msg01677.html

3 Text available at http://www.regjeringen.no/nb/dep/bld/pressesenter/pressemeldinger/2008/nordmenns-okologiske-fotavtrykk-er-malt-.html?id=504420

References

Ahmad, N. and Wyckoff, A. (2003). *Carbon Dioxide Emissions Embodied in International Trade of Goods*, DSTI/DOC(2003)15. Paris: Organisation for Economic Co-operation and Development (OECD).

Bastianoni, S., Pulselli, F.M. and Tiezzi, E. (2004). The problem of assigning responsibility for greenhouse gas emissions. *Ecological Economics*, **49**(3), 253–257.

Bruvoll, A. (2006). Future CO_2-emissions: a large increase in emissions from consumption. *ØkonomiskeAanalyser*, 5/2006, 25–30 (in Norwegian). http://www.ssb.no/emner/08/05/10/oa/200605/bruvoll.pdf.

Ecoinvent (2005). A database of life cycle inventories for nearly industrial processes, version 1.2. http://www.ecoinvent.ch.

Helm, D., Smale, R. and Phillips, J. (2007). *Too Good To Be True? The UK's Climate Change Record*. Oxford: New College, University of Oxford. http://www.dieterhelm.co.uk/publications/Carbon_record_2007.pdf.

Hille, J., Storm, H.N., Aall, C. and Sataøen, H.L. (2008). *Environmental Pressure from Norwegian Consumption and Production 1987–2007. A Report Commissioned by the Ministry of Environment and the Ministry of Child and Equity*, VF-raport 2/08, Sogndal, Vestlandsforsking (in Norwegian with English summary). http://www.vestforsk.no/www/download.do?id=814.

IPCC (1996). *Revised 1996 IPCC Guidelines for National Greenhouse Gas Inventories* (3 volumes). Geneva: Intergovernmental Panel on Climate Change.

Lenzena, M., Murraya, J., Sackb, F. and Wiedmannc, T. (2007). Shared producer and consumer responsibility – theory and practice. *Ecological Economics*, **61**(1), 27–42.

Munksgaard, J. and Pedersen, K.A. (2001). CO_2 accounts for open economies: producer or consumer responsibility? *Energy Policy*, **29**(4), 327–334.

Naturvardsverket (2008). *The Climate Impact of Consumption*, Report 5903. Stockholm, Naturvardsverket (in Swedish, with English summary). http://www.naturvardsverket.se/Documents/publikationer/978-91-620-5903-3.pdf.

NOU (2006). *A Climate Friendly Norway. An Exposition from a Government Committee Presented to the Ministry of Environment*, Official Norwegian Reports 2006:18 (In Norwegian) http://www.regjeringen.no/Rpub/NOU/20062006/018/PDFS/NOU2006 20060018000DDDPDFS.pdf.

Peters, G.P. and Hertwich, E.G. (2006). Pollution embodied in trade: the Norwegian case. *Global Environmental Change*, **16**(4), 379–387.

Peters, G.P.and Hertwich, E.G. (2008). Post-Kyoto greenhouse gas inventories: production versus consumption. *Climate Change*, **86**, 51–66.

SFT (2005). *National Inventory Report 2005 Norway: Greenhouse Gas Emissions 1990–2003 Reported According to the UNFCCC Guidelines*. Oslo: The Norwegian Pollution Control Authority.

WWF (2008). *Danish Consumption, Global Pollution. An Analysis of the Carbon Footprint of Denmark, with a Specific Focus on China*. Copenhagen, WWF (in Danish with English summary). http://www.wwf.dk/dk/Materiale/Files/Nyheder/Dansk+forbrug%2c+global+forurening.

6 Global warming and cultural/media articulations of emerging and contending social imaginaries

A critical realist perspective

Cheryl Frank

An overall framework – towards an emerging 'zyxa formation'?[1]

This chapter begins a more specific examination into the explanatory and transformative powers of critical realism, focusing on the problematic of climate change, and in particular on global warming. It will be argued that critical realism can gain in transformative power as a theory, methodology and indeed practice-informing approach to ameliorative action, by developing a theory of 'articulation', borrowed initially from British cultural studies, and also by becoming more aware of the symbolic and political power in how language operates through utilizing critical discourse analysis. The argument will first summarize the advantages of critical realism and then take the current perceived threat of global warming as the starting point to attempt to show, as a thought experiment initially, how we, as critical realists or something very close to it – through our schools, places of work, media, places of worship, secular organizational and institutional life, as well as in our wider cultural practices – may hope to transmit through our scientific and social-scientific research and other practices a more adequate understanding of the problem of global warming and what social action is needed.

It is argued that the comprehensive metatheory, and pluralistic, inclusive methodology of critical realism, makes it uniquely capable of helping to understand and co-ordinate human activity on the scale necessary to address the problem of global warming. Global warming, a highly complex and unevenly distributed, and changing, set of phenomena, is at once a material, discursive, political, moral and spiritual issue. The particular problem examined is the conjuncture of physical events, social structures, social interactions and individual practices loosely grouped together under the sign of 'global warming'.

Global warming is theorized as a range of phenomena that are multi-dimensional, multi-scalar and multi-faceted and that 'really' does exist within individual psyches, interactions between people, communities, ecological systems, larger societies and cultures, regions, nations and global expanses. It 'really' does exist in highly particular and complex ways on all levels of human existence, even on the smallest of communal levels, where people living simply and close to the earth may be directly affected and harmed, as for instance those living in villages on islands and along coastal areas such as the Maldives or even Bangladesh whose land is in danger of being submerged by rising ocean levels. Bhaskar propounds an ontology spanning cosmological through to everyday scales down to subatomic interactions. He shows how a theory of interdisciplinarity must address this expanse of things and events, powers and liabilities called global warming. Such a theory must seek understanding and transformation at the everyday moral, spiritual and historical conjuncture where human kind now precariously resides. This can only be achieved through unprecedented human intentionality and cooperation among people of the world.

The chapter focuses on how to make Bhaskar's substantial body of emancipatory political, moral and spiritual philosophy increasingly effective in explaining and intervening in urgent social problems. It is argued that Bhaskar's ontology, together with his concepts of 'four-planar social being', 'the necessity for interdisciplinarity', 'maximum inclusivity', the meaningfulness of the world *sui generis*, the grounding in reality of human solidarity, and the transcendental morality and reasoning of all human being provide philosophical stances which can begin to show the way in which this problem can be addressed and ameliorated.

This conjuncture called 'global warming' is also represented in various ways in the media and is articulated to other environmental, political, economic and social–cultural concerns as well. This historical conjuncture of global warming is real and serious enough, in that according to the best scientific assessment in 2009, the earth is actually warming to the point that life, i.e. the life forms together with the environmental conditions that sustain them, is threatened in the short term and possibly largely destroyed in the long term, at least on the worst scientifically and culturally emergent scenarios. These sober to catastrophic social imaginaries are projected alongside ones that are conflicting and contentious, imaginaries that range from somewhat skeptical to hopeful to celebratory. These counter imaginaries to the global warming thesis convey that the case for the overheating of the earth is not made; that the earth is not unduly threatened, or it is not warming because of human action but as a result, for example, of the Sun's increased radiation. Anyway, according to the anti-global warming imaginaries, there is nothing to do but let nature be nature and let the market be the market: as humans continue to act as producers and consumers, the unseen hand will presumably decide what to do about more and more areas of producing, reproducing and transforming life forms on Earth!

On the contrary, this chapter is concerned with the way in which critical realism produces grounds for life-affirming, emergent social imaginaries. Such emergent social imaginaries always have an element of novelty and unpredictability. Bhaskar has pointed out that big events such as the French Revolution and the demise of the Soviet Union cannot be generally anticipated, let along precisely predicted, or even often foreseen at all, although academic researchers and analysts, along with professional cultural and media interpreters, will often try to discern tendencies and directionality in social developments. There is hope that global warming can be ameliorated and that healthier ways of life can supplant the 'death culture' of the weapons-selling war industries and aggressive political regimes.

Concrete utopianism,[2] a Bhaskarian concept of progressive practice and hope, lets us imagine new social possibilities and organized co-operative modes of being. There is an everyday morality operating in even the smallest of human practices. Moreover, people generally know something of what they are doing and why they are doing it. A world that is sustainable and health-enhancing must become more of a real (possible) socially emergent imaginary. But it will be contending for hegemony with other imaginaries more rooted in ignorance, conspiracy, violence, war and terrorism, death and destruction – in other words, those pitting egos, groups and nations against each other.

This extensive body of work developed by Roy Bhaskar, beginning in 1975 with *A Realist Theory of Science*[3] – work which came to be called critical realism and now, since the turn of the century, has included the spiritual dimension, which he has called the philosophy of meta-Reality[4] – is just possibly uniquely capable of helping us understand and ameliorate our most profound challenge of surviving and flourishing along with other life on the planet Earth. Critical realism, as it has developed into a metatheoretical formation spanning over three decades, can help us to understand and coordinate human activity on the scale necessary to address complex social problems like how best to educate our children and how to help ameliorate global warming. Like the tremendous interdisciplinary, and indeed transdisciplinary, efforts needed to develop the atomic bomb, go to the moon, map the human genome, etc., the problem of global warming will also necessitate profound cooperative projects and determination at the highest levels through to the lowest to meet this challenge to life. This requires transforming and melding together the scientific and social scientific models of research, testing and application to make the overall scientific enterprise effectively address all levels of global warming, including physical, geo-physical, biological, socio-economic, socio-cultural, linguistic, and psychological dimensions.

Tools of the trade – the big picture

Critical realism provides a most comprehensive account, at the abstract level, of the 'real world' through Bhaskar's depth-ontology and his epistemological relativism, further developed in dialectical critical realism through the extended MELD system[5] and subsequently further deepened in his philosophy of meta-Reality. Bhaskar discusses and analyzes the many levels or dimensions of ontology: being as such and being as structure, being as process and change that incorporates absence and negativity and contradictions, being seen as involving totalities, being as transformative human agency, being as reflexive and spiritual, being as transcendence or 're-enchantment' of the world that is meaningful and valuable in itself, and all being as part of a wholeness or unity in diversity, with depth and variety which continues to unfold into new aspects. Another tool he has presented is a four-fold theory of truth[6] – that which is fiduciary and says 'trust me', that which is epistemically warranted through the best scientific and authoritative practice, that which is expressive as in 'the grass is green', and that which is alethic (a necessary causation as in emeralds are green because of their crystalline structure). Bhaskar explains social life in his system variously presented and known as TMSA (transformational model of social activity)[7] and 'four-planar social being' or 'the social cube',[8] where he gives an account of what happens in our social life. For everything we do, he says, there are implications on at least four levels: the level of material exchange with nature, the level of social structure, the level of social interactions between people, and the level of the person, which Bhaskar says consists in our higher selves or ground states (or souls, which can never be destroyed), our illusory egos (which make us think that we are separate from other people and must contend with them), and our real embodied personalities (with their mental and emotional experiences, including our subconscious or unconscious). More recently, Bhaskar has laid out an ontological theory of interdisciplinarity, of which more later. These basic tools are comprehensive, powerful and are being applied in an increasing number of disciplines and fields of study.

Critical realism's explanatory power widens, as many in academic life turn away from the ultra-relativistic, ultimately nihilistic, dead-ends of strong social constructionism, poststructuralism and postmodernism and the numbing effects of too much and often inappropriate number-crunching empiricism (as distinct from necessary empirical grounding). Critical realism takes an inclusive, pluralistic approach to methodology, so that the basic and very important insights gained through statistical analysis, surveys, ethnographies, social constructions and deconstructions are preserved and woven into our understandings of how the real world operates, including both the natural and social worlds. The Bhaskarian concept of judgmental rationality[9] allows us to choose which explanatory critiques have the most power to answer questions and solve problems, which ideology-critiques reveal the TINA[10] formations blocking us mentally or emotionally or hindering effective action, and which transformative critiques help make the world more just, leading to what Bhaskar calls a eudaimonistic society, a good society, where no person feels free unless all people are free.[11]

Enter the ground state[12]

At present, many people yearn to engage in corrective, transformative action, to do something effective and sustainable about this horrendous problem facing all humankind of global warming. In Bhaskarian terms, many people are increasingly experiencing the promotings of their higher selves, or spiritual or secular ground state qualities of care and concern for others, including other life-forms and ways of life. On an institutional level, schools and religious institutions are turning to ground-state qualities in their communications and actions to address global warming. Media and other organizations are campaigning for individuals and families to use less oil, gas, coal and electricity and give consumers more and more advice on how to down-size their carbon footprints, for example. In general, it may be said that global warming (within the context of overall climate change) has been put on the media agenda. But how it is framed is something we need to look at more closely. More later on this.

Shedding the 'natural attitude'

For quite some time – at least from the time of Descartes and the onset of modernity – until about the last half of the twentieth century – the environment, then called 'nature', was largely a 'given' for the overwhelming majority of people, something separate from human beings, and to be manipulated and operated on by them, though there were a few prescient scientists such as Tyndall and Arrhenius who were concerned with the effect of changes in carbon dioxide concentrations on climate as early as the second half of the nineteenth century.[13] It is true that there had been mounting worries, on the part of the Romantics and others, about the blight of growing industrialization in some cities and geographical areas, especially in the West.[14] However, the environment was conceived as part of received reality and was represented in language as beautiful or sublime nature such as awesome mountains or glorious scenery. Moreover, nature was integral to life in the West, at least, to wider cultural practices, like holidaying at the seaside or walking and hiking or picnicking, or boating on the lake, or indeed travelling abroad to enjoy distant lands. What people said and thought about the environment was not, for most, part of a contentious, contradictory discourse. In the common view, the environment remained a positive presence, with few worrisome negatives or absences – except for occasional catastrophic 'natural' events such as the Lisbon earthquake or for the acknowledged sublime powers of the elements. In general, there were few gaps or dangers or particular fragilities, at least not any that seemed out of the ordinary or insurmountable. Certainly very few (except perhaps for some prescient scientists, nature worshippers, end-world thinkers, more pessimistic social analysts such as Malthus and Darwin, and science-fiction writers and dedicated hypersensitives) thought about the possibility that agricultural, industrial and consumption patterns could create a problem of global warming or indeed species extinction and destruction of all life as we know it. Most people went about their lives and business in what Bhaskar calls 'the natural attitude', that nature was separate from humans, to be used and enjoyed by them, and overall virtually indestructible – something that was just out there.

Then the situation began to change; there began an absenting, in Bhaskarian terms, of this naturalistic attitude, as evidence began to mount up, and normal science came to appreciate, that not all was right with the world. A paradigm shift was under way. For years, some scientists fought among themselves over whether global warming and other climate change was really happening. And certainly some scientists squared off against environmental activists and educators over whether the earth was heating up, or if it was, whether the collective action of humankind was the major cause, and if it was, whether anything could be done about it in time to save the planet. Now, people are no longer ensconced in a natural attitude about the environment where ontology (being) and epistemology (knowing) seem as one non-problemmatic, non-contradictory thing. Whereas understanding about the environment had once seemed natural and unquestioning, now people associate the environment with global warming and threatening climate change (on the ontological level) and with political squabbling and contentious discourses and accounts of the heating up phenomena (on an epistemological level). Further, people now realize that the environment is also fragile and actually may be destroyed or self-destruct, so that all life, including humanity, could, astoundingly, actually perish.

Anyone who reads newspapers and listens to other media accounts, will probably associate global warming with a growing number of other environmental problems as well (some of which are contradictory). These other problems include, for example, increased suffering and dying from rising temperatures; dwindling rainforests, desertification of land and increased forest fires; growing world hunger and the need for a new 'green revolution' to feed people as crops fail; how the old 'green revolution' in the last half of the twentieth century helped to cause many of today's pollution and land problems; the need for more effective agricultural inputs such as pesticides and fertilizers (or not); how GMO (genetically modified organisms) pesticides and foods are needed (or not); how GMO self-destructing seeds will increase production whether for good or ill; how such GMO seeds will drive even more independent farmers in developing countries out of business because of loss of control over their supply of seeds; the scarcity of water and increasingly depleted aquifers and underground water supplies; the rising of ocean levels and the drying up of lakes and reservoirs; the possible shut-down of the Gulf Stream conveyor belt; seriously rising numbers of plant and animal species going extinct or under dire threat; increased biofuel production to lower carbon emissions or, how to the contrary, biofuels are adding to the problem of carbon emission; the need to drill for more oil in formerly protected wildlife or scenic areas (or not); the need to build more nuclear power plants to replace carbon-based fuel (or not); how we need to develop solar, wind and so-called alternative energy sources; how wind farms may be spoiling the natural beauty of landscapes or seascapes; how increased consumerism and industrial production in the developing world – such as in China, India and Brazil – must be slowed or controlled (or not); how advanced capitalist nations must bear the brunt of sacrifice to ameliorate global warming (or not); and the remaining plethora of issues that have been framed as part of the global warming complex.

People no longer think that if there are still arguments going on over the environment, and, of course, there are, and whether and how much the globe is heating up and to what extent this threatens, or not, all life on Earth, that this should overrule what they perceive through their judgmental rationality as the best consensus of scientific understanding. Most people have made the paradigm shift in their attitude toward the environment and global warming. They believe that the earth exists independently of humans, but it also exists in interaction with humans, and is now severely threatened; that many species of plants and animals have been dying out for some time; and that many others are near extinction. The first hurdle of changing people's minds about the seriousness of the problem seems to have been overcome, not everywhere, not on all environmental vectors, but at least substantially in the West, which is being addressed in this chapter.

Absenting powerless to act – critical realism to the fore

Now there are other hurdles, such as the absenting of the feeling many have that it is too late to do anything – because power, politics and profit will prevent effective action and large corporations, together with the WTO and World Bank and other capitalist-friendly institutions, will continue to run the show by using and funding unsustainable practices.

The media campaigns, too individualistic, too much geared toward people and households, are not enough – people need the help of their governments and locally tailored councils and villages and policies, in addition to major institutions, to make the kind of headway needed. Schools, places of worship, universities, media organizations, voluntary organizations, environmentally concerned groups of citizens, political organizations and policies, concerned businesses – all must empower people to access information and learn for themselves how to understand scientific knowledge about global warming and the social problems and stakes, leading to their own, creative and corrective actions. There must be a groundswell of movement from below, which can be greatly aided by policies and empowerment from higher levels of collective organization and agencies. Can critical realism help overcome this powerlessness of people's feelings in this effort?

There is hope and a real possibility that this kind of comprehensive understanding, informed by critical realism, or something very close to it, could have major effects in transforming the social structures and social interactions that perpetuate global warming and drastic climate change. Some critical realists such as Petter Næss, of Aalborg University in Denmark and Oslo University College in Norway, Karl Georg Høyer, also of Oslo University College, Hugh and Maria Inês Lacey who are academics and activists in the USA and Brazil, Jenneth Parker and Sarah Cornell, both of Bristol University, and certainly many others, already devote all or some substantial portion of their academic work directly to a variety of environmentalist concerns such as sustainable urban planning in Næss's case, organizing Norway's scientists and activists in Høyer's case, working with interdisciplinary scientific teams in Cornell's and Parker's case. Many critical realists, or those close to this position, are teaching and writing or are otherwise politically engaged in issues pertaining to the environment in related fields. Næss has written persuasively about how capitalism itself, as an economic system and way of life, is not sustainable and must be transformed.[15] So, critical realists as a whole and as individuals, if so moved, and those who adhere to the philosophy of meta-Reality in Bhaskarian thought, have a big challenge: if global warming is as urgent as we think it is, what can we all do to get this issue before the public in the various ways we know of, to reach and persuade people to take action.

One potentially very significant model of action has recently been announced in the UK in a *Guardian* article, banner-headlined on the front page, 'From the melting frontline, a chilling view of a warming world'. The subheadline reads, 'Public figures and business to sign up to climate drive'.[16] Correspondent David Adam writes,

> An unprecedented coalition of scientists, companies, celebrities, local councils and organizations spanning the cultural and political spectrum will today commit to slashing their carbon emissions as part of an ambitious campaign to tackle global warming.
>
> The 10:10 campaign, which will be launched at London's Tate Modern this afternoon [September 1, 2009], aims to bolster grassroots support for tough action against warming ahead of the key global summit in Copenhagen in December.
>
> Those signing up for the campaign, with is supported by the *Guardian*, pledge to make efforts to reduce their carbon footprints by 10% during the year 2010.

Here it is suggested that coalitions such as the 10:10 campaign – which itself ranges in membership from the *Guardian*, Tottenham Hotspur football club, an online grocery, the aforementioned world-famous contemporary art museum, the Women's Institute, dozens of schools and universities, NHS health trusts and several energy companies – need to be encouraged, supported, funded and organized on all levels of society to address particular problems or sets of problems in a coherent fashion. This would be done by intentionally marrying relevant laminated systems, where as many of the necessary elements of a particular problem or problem set on all levels are identified and addressed[17] – to those elements that are culturally relevant agencies, institutions, media and entertainment products and events, interventions in relevant discourses and scientific theories, etc. The aim would be to address the problem more fully in what I will call here an *intentional ameliorative and articulated laminated system*. This means that the objective agencies, structures, events, and powers that exist to perpetuate a problem will be further intentionally laminated also at the relevant social–economic, socio-cultural, psychological and moral levels as well. Citizens, groups, institutions and governments can then be 'joined up' to help solve the problem concerned, rather than working at odds with one another or not acting at all to help ameliorate the problem. This needs a 'double articulation'. First, of our understanding of the structure of the problem in an explanatory laminated system, and second, as our analysis of the indispensable units for comprehending and tackling the problem, involving the relevant agencies, organizations, discourses and resources to enable effective intervention to ameliorate the situation at hand.

This process would involve enlisting the social–imaginary, concrete utopian path or solution which Bhaskar has described. It would empower or resource peoples' ground states (or higher selves) and enable more effective forms of human agency to address our most pressing problems. Critical realism provides the tools – the philosophy, method and vision to help forge the way in this regard.

To this end, some salient advantages of critical realism have already been briefly noted, and before proceeding to British cultural studies and articulation, it is well to emphasize other advantages of critical realism. A great advantage, as elaborated by Bhaskar, partly emanates from what will be called here the ontological theory of the necessity of interdisciplinarity.[18] Another advantage in critical realism stems from its maximally inclusive principle of acknowledging the value of findings arrived at by different systematic methods of study in the social and human sciences, including empiricism, Neo-Kantianism, hermeneutics, poststructuralism and postmodernism.

The linking of these three elements, (1) interdisciplinarity, (2) maximal inclusivity and (3) British cultural studies articulation theory and practice, involves, or it may be argued, even necessitates, the concept of 'dialectical articulation'. The concept of articulation referred to above follows and builds upon the thought of Stuart Hall, a leading sociologist and cultural theorist who came to Britain from his homeland Jamaica. He directed the Centre for Contemporary Cultural Studies at Birmingham University from 1968 to 1979, having taken over the helm from Richard Hoggart.

Articulation theory and British cultural studies to the fore

The idea in British cultural studies of 'articulation'[19] is an additional powerful tool at the critical realist's disposal that can be used in connection with analyzing culture. Culture is seen as a mutli-faceted concept that can mean many very different things, depending on the context. Culture usually means in British cultural studies, following Raymond Williams, a whole way of life, or a structure of feeling and thought.[20] Hall stressed the practices and products and emotions of everyday life and how people appropriate elements of culture and transform them. He and his colleagues discussed what they termed 'the circuit of culture', and

> the articulation of production and consumption. In this model, cultural meaning is produced and embedded at each level of the circuit. The meaningful work of each level is necessary, but not sufficient for or determining of, the next instance in the circuit. Each moment – production, representation, identity, consumption and regulation – involves the production of meaning which is articulated, linked with, the next moment. However, it does not mean determine what meanings will be taken up or produced at that level.[21]

So, for example, a product, such as a Sony Walkman or a newspaper story about global warming, can be analyzed as to what the designer and producer or editor and reporter intends it to mean (or what it represents); what the advertiser or perhaps television announcer says it means or represents; and what the product means to those who purchase the Walkman or particular newspaper or television program and, if different, to the actual consumers, users or readers or viewers, of it. The main point is that the designers and producers such as electronic manufacturing corporations (in the case of the Walkman) and newspaper owners and managers and producers such as editors and reporters (in the case of the story about global warming) do not directly determine what meanings the consumer associates with the product and how the product may or may not become part of the consumer's identity. (By the way, this seems to be in some tension with a kind of top-down model of 'manufacturing consent', such as Noam Chomsky's, who is greatly to be admired for his theoretical and activist work, but may have a too simplistic model of how hegemony is in fact always to be won and negotiated in a particular context such as national politics.) A model which shows how people are capable of acting on information and narratives and bringing their own experience and meaning to bear seems to more accurately reflect the 'messiness' of cultural meaning-making and everyday life; and it also seems to give more hope for the possibility of corrective action and change.

This kind of thinking led Hall and his associates to develop a theory of reception and encoding/decoding that essentially criticizes the idea that people are 'cultural dopes' who just take dominant media and political messages and symbols as given. According to this school of cultural analysis, people take up messages and products from, for example, television, newspapers and other media, and often make their own meanings from them. Of course, people may 'decode' from a product what the manufacturer and advertiser 'encode', but they may not. People, according to this cultural model, appropriate embedded signs, stories, words, and affects and fit them into their own ways of thinking, understandings, feelings, coping strategies, identities, subjectivities, and intentionalities and moralities. This is in line with critical realist thinkers such as Bhaskar[22] and Margaret Archer[23] who contend that people and organized social groupings of people in their agential capacities generally know something of what they are doing and why they are doing it, and with the thought of Andrew Sayer,[24] who sees an everyday basic reciprocal and consistent standard of morality and human dignity permeating all aspects of the everyday life world – in other words, people generally want to do the right thing, treat each other well, and bring up and educate their children within a socially acceptable moral standard.

To continue the immediate discussion: Hall showed how things, events and feelings get linked in the circuit of culture and perhaps the even wider cultural context through a process of what he called *articulation*. Chris Barker paraphrases Hall as saying society is constituted by a set of complex practices with their respective specificities and modes of articulation standing in an 'uneven development' to other related practices. Hall famously likened this process of articulation to the metaphor of a lorry. He described hooking up a lorry, where the cab can be over time articulated or connected to any number of chassis, showing there is no necessary connection between a particular cab and a particular chassis. It is human (usually) emotional identity, along with other sets of structures and constraints, that brings the two together in a particular context in the open social world where natural necessity generally is not a factor in the same way it is on the physical and perhaps biological level (which does, to one extent or another, interact with the social world, of course). But this metaphor is intended to show how articulation works in discourse and popular culture: in other words, things are often linked together, usually not by necessity, or logic or even wholly rational considerations for the most part, but very often because of their affective powers to persuade and manipulate consumers and citizens.

This persuasion can come through political hegemonic rhetoric from above (such as from politicians, mainstream television and other media), with their generally dominant ways of thinking, which could be classified as ideologies if they keep some people in power and wealth and others in ignorance and oppressive conditions such as poverty or ill-health. In contrast, things may be linked together from below by a process of cultural uptake and pleasure such as when military clothes are taken out of context and worn by street dancers to signify something new and alternative; or when young people strive to forge new identities through production and consumption of music; or through protracted political struggle, as when an alternative or counter-hegemonic discourse struggles to gain ascendency over another.

So that, for example, in Britain and American following 9/11 and the outrage and bewilderment many people felt, public opinion was vulnerable to massive manipulation. Over the next few years, the majority public opinion in the US went from largely supporting an irrational, unjust war against Iraq at its inception to in the last few years apparently opposing it. This of course did involve discourse and cultural construction of events and meanings, but it was also 'real' in the sense that a war did happen, people died and were injured and bereaved. People at first were led to believe the war was just and necessary because certain things were articulated together. You will recall we were told by politicians that Iraq was the home of a madman, the dictator Saddam Hussein, who was one of the worst heads of state because he tortured and gassed dissenters and Kurdish people, that somehow Iraq was the central repository of support for Osama Bin Laden and Al Quaeda, that Iraq had weapons of mass destruction, that hard evidence of these weapons had been seen by surveillance satellite, that some of the WMD were capable of striking Britain within 45 minutes, etc. These things were articulated together over and over by politicians, government officials and the military speaking through the mass media, until people accepted that war had to be waged against not only Iraq's military but its civilian population.

Now we have disarticulation of these things (most were unjustified or plainly a lie) and a rearticulation that the war was fought for oil, to maintain US hegemony of power in the world, to enrich certain major corporations, to line the pockets of certain politicians, etc.

The context of 9/11 and the alarm people felt, exacerbated by the onset of the 'War on Terror', was a highly combustible issue – perhaps like global warming. For global warming is an issue that will probably become more pressing, more alarming, more urgent. This means that articulations and rearticulations are being made about what causes it, what to do or not to do, what countries, industries, organizations, schools, families, individuals should do about it. Global warming therefore needs to be theorized as a range of articulated, disarticulated and rearticulated phenomena that are multi-dimensional, multi-scalar and multi-faceted and which are both discursive and non-discursive. Late capitalism in some places may be directly articulated to global warming, but in other places, it may not be.

Emergent social imaginaries through concrete utopian thinking and eudaimonic practices

As already remarked, emergent social imaginaries always have an element of novelty and unpredictability. Big events such as revolutions cannot be predicted by social science in any specificity for good reasons – they occur in open systems characterized by emergent qualities and human creativity. This means that there is always possibility and novelty for effective intentional action, and hope that global warming can be ameliorated.

As noted, concrete utopianism lets us imagine new social possibilities and organizationally cooperative modes of being. We can look at the vast resources we have and imagine how they could best be used to better human life and end suffering. Such thought experiments can, if widely articulated, become part of an ascendant discourse and enter wider realms of public and professional consciousness. There is, to remind us, an everyday morality operating in even the smallest of human practices, according to Bhaskar and Sayer. Moreover, a world that is sustainable and enhancing must become more of a real possibility. But it will be contending for hegemony with other imaginaries of ignorance and oppression, death and destruction. Gramsci's model of struggling for hegemony, in conjunction with Hall's idea of articulation applied more widely, can be seen to pertain to how we feel, act and think in our everyday lives, as well as how we participate in research agendas, what kinds of questions we ask and how they are framed, as well as how we relate to the mass media and to the political realm. Critical realism and meta-Reality, through its calling to our highest selves and the emancipatory impulse, can mobilize people and resources to articulate a greater, transcendent totality of how to organize and act.

Critical discourse analysis to the fore

Norman Fairclough has written that he studies 'the place of language in social relations of power and ideology, and how language figures in processes of social change'.[25] In his book *Language and Globalization*,[26] Fairclough shows how critical discourse analysis can help us understand the properties and features of words, art, representations, texts, language and signs (discourse) that bear the styles and forms and genres that carry the meanings which many critical realist researchers may want to consider appropriating into their understanding of how the increasingly mediatized social world operates. This approach to discourse analysis, Fairclough maintains, works especially well in interdisciplinary projects. Fairclough works closely with critical realist Bob Jessop who writes from a perspective he calls cultural political economy,[27] recognizing that language and culture are necessary to explaining phenomena in the field of political economy. Jessop also theorizes the so-called knowledge-based economy and how the rhetoric behind it serves political ends. Much has been written by Fairclough, Jessop and others about 'globalization': there is a growing discourse of explaining, criticizing and contending with problems of globalization, from accounts to the history of world trading and political systems; to the idea of the destructive practices of institutions with global reach, such as the World Trade Organization, the World Bank, the International Monetary Fund, and the Multinational Corporations; to the idea of an evolving global, one-world government; to how globalization affects trade and local production and consumption; to the idea that people are becoming more mobile and are forming a 'world citizenry'; to the idea of hegemonic and monolithic global capitalism as inevitable. This last example may be a TINA (There Is No Alternative) formation, in Bhaskarian terms. Are we therefore to teach people that oppressive capitalism is growing and inevitable? Not at all. We must remind people that there is always novelty in the world and hope for transforming it and that human beings, through their intentional agency, can make a crucial difference.[28]

Critical discourse analysis uses the concept of intertextuality, showing how discourses are always in dialogue with one another, that virtually any story or account will incorporate the language and meanings of previous texts and give them variously different inflections. These intertextual discourses are also very often in contention, trying to win the hearts and minds of people, some even contending ultimately, by certifying their articulations as knowledge, for purposes of maintaining, contending or winning hegemony. Western societies are contending for hegemony in the area of global warming, now that the natural attitude is pretty much abandoned. But what new discursive ways of thinking will take its place? The issue of global warming will often be addressed through genres of news media, such as the investigative report, the hard news story or narrative; the trend story, the interview, the editorial, the feature, the consumer opinion story, etc. Such stories, almost always overtly or covertly political, will very often adopt a government frame or a frame being urged by powerful interests. If the story (text) is a national political story, and particularly if only few or mostly marginalized national elites are contending for what will become or remain the dominant frame, it is likely a government frame will be adopted. So, in the absence of other frames, media will often adopt their government's frame of telling the story, by positioning certain actors or organizations as more right or morally correct than others. This helps to understand how governments can at first manipulate public opinion relatively easily.

Bringing local knowledge back in, enhanced by science and used with wisdom

Much local knowledge – and wisdom on stewardship of the environment and how to 'tread lightly on the earth' (as Native Americans often say in the USA) — is now mostly submerged in present dominant cultural significations and practices. Local knowledge must be allowed to be used and assessed according to our judgemental rationality and our most disinterested scientific understandings. But local traditional and current popular knowledge around the world can be fostered and brought more to the fore and articulated in the process of widening popular culture for various public spheres, as knowledge is received, (re)contextualized and (re)appropriated according to the local needs of people and the environment. Scientific knowledge must also be fed into the stream of knowledge circulating in media, culture, places of worship, political arenas, schools, and everyday life. There must be a much more effective media feedback loop and enrichment of wisdom between and among peoples, groups, institutions, science and maybe even especially including various modes of popular entertainment. Bhaskar wants more of us to be in touch with our ground states or our most basic and best natures, or higher selves, which no one can really lose because it is part of being human, no matter how conflicted or split or deranged a person becomes. As more people get in touch with their higher selves, they want to act cooperatively to ensure the earth, and the life it supports, endures. Political practice must tap into this rich source of moral yearning and empower it.

To this end, local knowledge needs to be articulated with school knowledge, cultural understandings, media productions, and everyday morality in newer, much more progressive circles of culture. Here it is being suggested that by adding articulation theory from cultural studies and critical discourse analysis to the developing body of critical realist theory, the researcher and activist will be able to address the concerns of global warming and climate change generally in more effective ways. These concerns include arenas of academic research and political action, designed to percolate and spread transformative social change. Perhaps we will always be learning how to build eudaimonia (the good society or at least the best possible one now) through zyxa formations at every level of the social, where there is growing persistence of optimism of the will and growing power of realism of the intellect.

Notes

1 This chapter is launched with reference to the light-hearted, but still philosophically serious, concept of 'zyxa'. Invented by Mervyn Hartwig, the editor of *Journal of Critical Realism*, the concept is based on a phrase from Roy Bhaskar. It is the idea that, contra to Gramsci, one does not need optimism of the will and pessimism of the intellect, but rather, 'optimism of the will and *realism*, informed by concrete utopianism, not pessimism, of the intellect' [quoted from R. Bhaskar, *Plato Etc.*, 1994/2009 2nd edn, Routledge, London, p. 215], such that, as Hartwig puts it, 'freedom is won'. Hartwig coins the word zyxa in his *Dictionary of Critical Realism*, Routledge, London, 2007, p. 503. This dictionary, henceforth abbreviated as *Dictionary*, will be used throughout

to refer those interested in deeper exploration and references related to critical realism.

2 Dictionary, pp. 74–75.
3 Roy Bhaskar, *A Realist Theory of Science*, 1975/2008, 4th edn, Routledge, London.
4 See, e.g. Roy Bhaskar, *Reflections on meta-Reality*, Sage, London, 2002.
5 *Dictionary*, pp. 295–303.
6 *Dictionary*, pp. 484–488.
7 *Dictionary*, pp. 467–470.
8 *Dictionary*, pp. 420–1.
9 Judgemental rationality, along with epistemic relativism and ontological realism, comprise 'the holy trinity' of critical realism. See entry on 'holy trinity' in *Dictionary*, pp. 238–42.
10 TINA stands for 'there is no alternative', a slogan used by former British prime minister Margaret Thatcher, and appropriated and redefined and recontextualized by Bhaskar in his analysis of irrealism. See *Dictionary*, discussion on 'Tina syndrome', pp. 465–467. Bhaskar says there are indeed alternatives to TINA formations, i.e. that there are alternatives available, for example, to those who say nothing can be done about war or ameliorating global warming, or to rapacious, unsustainable capitalism.
11 *Dictionary*, p. 187.
12 The term 'ground state' is a term Bhaskar uses to include not only the more religious concept of 'soul' but also the secular concept of our best or 'higher selves'; all ground states are connected up in what he calls the 'cosmic envelope'.
13 See Karl Georg Høyer's Chapter 3 in this volume, and also Chapter 7 by Sarah Cornell.
14 For those interested in a very close historical analysis of how nature and the environment have been represented, see David Pepper, *Modern Environmentalism: An Introduction*, Routledge, London, 1996.
15 Næss, Petter, Unsustainable growth, unsustainable capitalism, *Journal of Critical Realism*, 5.2, 2006.
16 *The Guardian*, Tuesday, 1 September 2009, pp. 1–2, written by David Adam, environment correspondent.
17 The term 'laminated system' was employed by Roy Bhaskar and Berth Danermark in their article, 'Metatheory, interdisciplinarity and disability research: a critical realist perspective', *Scandinavian Journal of Disability Research*, 8(4)(2006). For further elaboration, see Chapter 1 of this book. In general, something like 'articulated laminated systems' must be identified as the indispensable units for comprehending and tackling complex, multi-level phenomena such as climate change.
18 The ontological necessity for interdisciplinarity was first spelled out in detail by Bhaskar in the article he co-authored with Berth Danermark – see Bhaskar and Danermark, *op. cit*.
19 Articulation means: (i) a temporary unity of discursive elements that do not have to 'go together'; (ii) the form of connection that *can* make a unity of two different elements, under certain conditions; (iii) expressing/representing; and (iv) a joining together. Reformatted and quoted from four bullet points in Barker, Chris, *Cultural Studies: Theory and Practice*, 3rd edn (with a foreword by Paul Willis), (Sage, London, Thousand Oaks, California, New Delhi, and Singapore), 2008, p. 59.
20 Williams, Raymond, *Keywords*, 1973/1988, Harper Collins, London.
21 See Barker, *op. cit*., p. 59.
22 Roy Bhaskar, *The Possibility of Naturalism*, 1979/1998, 3rd edn, Routledge, London.
23 Margaret Archer, *Making Our Way About the World*, Cambridge University Press, Cambridge, 2006.
24 Andrew Sayer, *The Moral Significance of Class*, Cambridge University Press, Cambridge, 2005.
25 Norman Fairclough's website at Lancaster University: http://www.ling.lancs.ac.uk/profiles/263

26 Norman Fairclough, *Language and Globalization*, Routledge, London, 2006.
27 Bob Jessop, *The Future of the Capitalist State*, Polity, Cambridge, 2002.
28 This point has been made by critical realists, for example, Jenneth Parker, 'The theorization of collective ethical action', Chapter 7 of her unpublished Ph.D. thesis.

7 Climate change

Brokering interdisciplinarity across the physical and social sciences

Sarah Cornell

Major investments have been made in research programmes worldwide that try to bring physical and social sciences relevant to climate into positive working relationships. This chapter will review the ways in which knowledge integration has developed over time across the physical sciences in relation to climate change, and the rationale and issues of joint working that have resulted. It will then proceed to outline some of the challenges of developing meaningful and useful interfaces between this somewhat interdisciplinary group of physical sciences and the social sciences. A critical perspective will be adopted in terms of the theoretical and philosophical underpinnings or enabling frameworks for this project. This latter section will explore the potential of critical realist approaches to interdisciplinarity and will raise questions towards an expanded research agenda for critical realism as a philosophical research programme.

The foundations of 'climate science'

Although climate science as a distinct field of enquiry is really only about 50 years old, it has been built up over a very long time from interdisciplinary foundations. Its parent subjects – meteorology and oceanography – have a much longer pedigree. Humans have acquired knowledge of the dynamics of the atmosphere and oceans since pre-historic times, but it was in the eighteenth century, a period of overlap between the Enlightenment and the era of merchant voyages, that these studies started to be documented systematically. As an aside, these early climate scholars would have regarded themselves as 'natural philosophers'; unlike today's specialist scientists, most demonstrated a certain fluidity in their investigations of both the wet and the windy milieux, and many were polymaths investigating many other natural phenomena too. Theirs was 'normal' science – perhaps not strictly in the Kuhnian sense of uncritical operation within a paradigm, but nevertheless based on hypotheses generated from theories conceived to explain observations, and in turn tested experimentally against observations. Arguably, their pre-disciplinary efforts were part of the process of the deepening and narrowing specialization of (empirical) scientific knowledge that constrains us so awkwardly now. Later in this chapter, I suggest we are currently following a cycle back to an ideal of full knowledge integration and adisciplinarity, the 'romantic knowledge ideal' of our own times.

Meteorology became a modern science, rather than lore, with the proliferation and improved precision of instruments such as barometers and thermometers developed in the seventeenth century. Theoretical understanding developed of tides, winds and climate systems (such as the seasonal monsoons), which required the conceptual connection of weather and climate to the behaviour of the sun, the oceans, and indeed to the cosmos. It also became institutionalized, with the formation of the International Meteorological Organization in 1873. By that time, many countries had national meteorological services, and the newly formed intergovernmental organization was set up to be a specialized agency for weather and climate science.

Oceanography was developing in parallel. In the English-speaking world, Robert Boyle, the chemist, analyzed and documented the 'saltness' of the sea in 1673, linking it to weather variability and discussing the stratification of salinity in ocean waters. James Rennell (who in today's terms, would be labelled a geographer, historian and hydrographer) wrote about the prevailing currents in the Atlantic and Indian oceans in the 1770s, an early systematic effort towards oceanography. The Challenger (1872–76) and Discovery (1901–04) expeditions were major British efforts in empirical oceanography, and Norway, Germany and the United States of America were also exploring and documenting the world's oceans. Oceanographers also formalized their discipline, setting up the International Oceanographic Commission in 1960 to promote international co-operation in research and the protection of the oceans.

Climate science is now most strongly associated with the codification of understanding in global models. Numerical modelling as a tool emerged with the first efforts at weather prediction in the early 1950s. 1955 saw the development of the first atmospheric general circulation model (GCM), which simulated the fluid motion of the Earth's atmosphere to calculate winds, heat transfer and rainfall. Similar models were developed for global ocean circulation. Although it is quite amazing now to think of those early models being used to *predict* the consequences of human alteration of atmospheric chemistry, the first study of the effects of doubled atmospheric carbon dioxide concentration, by Syukuro Manabe and Richard Wetherald, was back in 1975, when the prospect of reaching an atmospheric CO_2 concentration of 380 parts per million (ppm) by the year 2000 still seemed remote (we are currently at 387 ppm, rising at around 2 ppm each year). An image published in the Intergovernmental Panel on Climate Change's First Assessment (IPCC 1990/2; Figure 7.1) illustrates the mechanism for the evolution of climate change science: models of different processes, such as atmospheric chemistry, ice dynamics, the seasonality of vegetation and so on, are developed offline and are progressively 'coupled' to the main model. Inasmuch as these offline models are developed within their specialist disciplines (chemistry, plant physiology and so on), drawing on the knowledge developed within those diverse research communities, climate modelling can itself be regarded as an interdisciplinary enterprise. However, throughout their incremental development

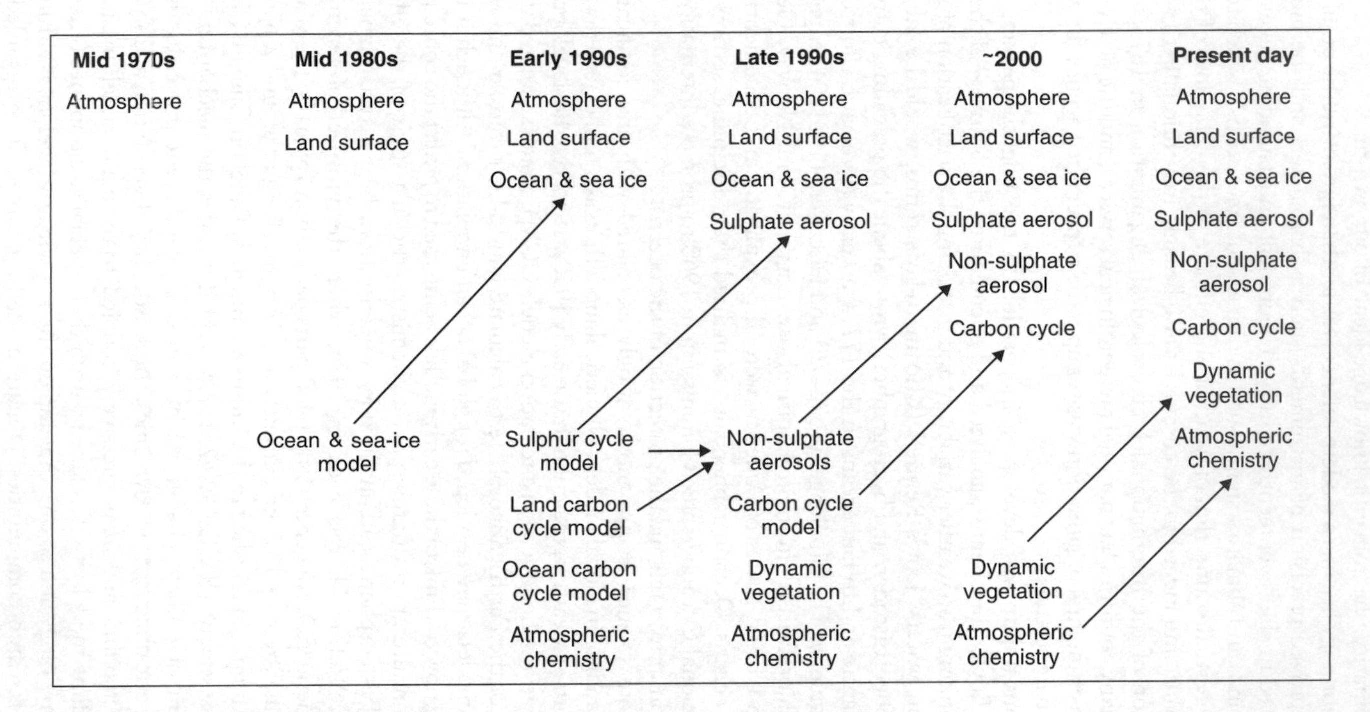

Figure 7.1 Development of climate models (based on IPCC 1990/1992)

pathway, climate models have been focused on representing and simulating Earth's physical processes, and this research has been solidly in the domain of the physical sciences.

One important interface that has arisen in global climate science over the past 20 or 30 years is that between the theoretical/conceptual and the observational/empirical; in many ways, this is an example of the challenge of an interdisciplinary interface *within* a field of science. Evans and Marvin (2004) call this 'cognate interdisciplinarity'. In 1975, the first weather satellite was launched. This satellite, GOES (the first in NASA's ongoing Geostationary Operational Environmental Satellite programme) enabled comparisons of global model outputs and Earth observations, with the promise of model 'validation' (assessment of the goodness of fit of the model to the data), in principle at least. However, model validation is still an under-developed area, in part because the knowledge communities dealing with model development and with empirical measurement and observational data tend to be separate, each with their own distinctive practices and values. A recent response to this problem is the launch of the international journal *Geoscientific Model Development* (Annan *et al.*, 2008), which creates a forum for benchmarking and critical review of models. There are other concerns about models and their relation to the actual world (e.g. Oreskes, 1994) that might benefit from a critical realist consideration; I shall return to these concerns in the following section.

Earth system science is sometimes presented as a Kuhnian paradigm shift from the empiricist, positivist, determinist framings of climate science (Lenton, 2002), but both the physics roots and the ethos of 'old' climate science are still very evident in the 'new' Earth system science. In fact, these framings developed together, and are closely interlinked. NASA's Mariner expeditions to Venus and Mars through the 1960s provoked new questions about the interplay of climate, atmospheric chemical composition, and life. The idea of Gaia emerged around the same time (Lovelock, 1972, 1979). Lovelock's hypothesis about Earth's feedbacks was that the living and non-living components of Earth interact to maintain climate and biogeochemical cycles in homeostasis, a self-regulated stable condition. Lovelock's 'new age terminology' – really little more than his choice of the earth goddess' name Gaia for the idea – and the suggestion of teleology in his initial colourful writings about the Gaia hypothesis were contentious at the time, but the science itself was not. (The teleology relates to the idea that climate is not just regulated by the biosphere, but that the biosphere somehow purposefully plans climate optimization for itself.) Earth system models, representing the feedbacks and interconnections of the subsystems, have developed from the original climate models with neither a conceptual shift or break nor any reconfiguration.

Understanding Gaia scientifically – identifying and quantifying global-scale processes, budgets of vital elements like carbon, nitrogen and silica, and Earth system feedbacks – was part of the motivation for the creation in the mid 1980s of a series of international Global Change Programmes, sponsored by the International Council for Science together with the World Meteorological

Organization and the International Oceanographic Commission. These global collaborative research programmes all recognized the need for new interdisciplinary working. They took the then radical steps of linking biology, geophysical processes and chemistry (the International Geosphere–Biosphere Programme, www.igbp.net; NASA, 1986; Figure 7.2) and of attempting global-scale synthesis, for instance, building on the year-long Global Weather Experiment of 1979 with the World Climate Research Programme's Global Energy and Water Cycle Experiment. In recent years, these Global Change Programmes have united under the aegis of the Earth System Science Partnership. More information can be found on the projects and their history on www.essp.org.

There is a further dimension of interdisciplinarity in present-day climate research. Understanding of past climates is increasingly being sought in order to inform understanding of the present situation. The historic record of climate and past landscapes is collated from palaeodata sources as diverse as counts of pollen and charcoal grains in lake sediments, the ratio of trace metals in the shells or skeletons of marine plankton, and fossil air bubbles in ice cores. These reconstructed pasts will, it is hoped, constrain the answers to questions like: (when) will there be another ice age? How hot could Earth get with a doubling of CO_2? What would this mean for forests and other land ecosystems? What happens to carbonate sediments when a major pulse of CO_2 is put into the atmosphere, and what does this tell us about the prospects for today's marine organisms?

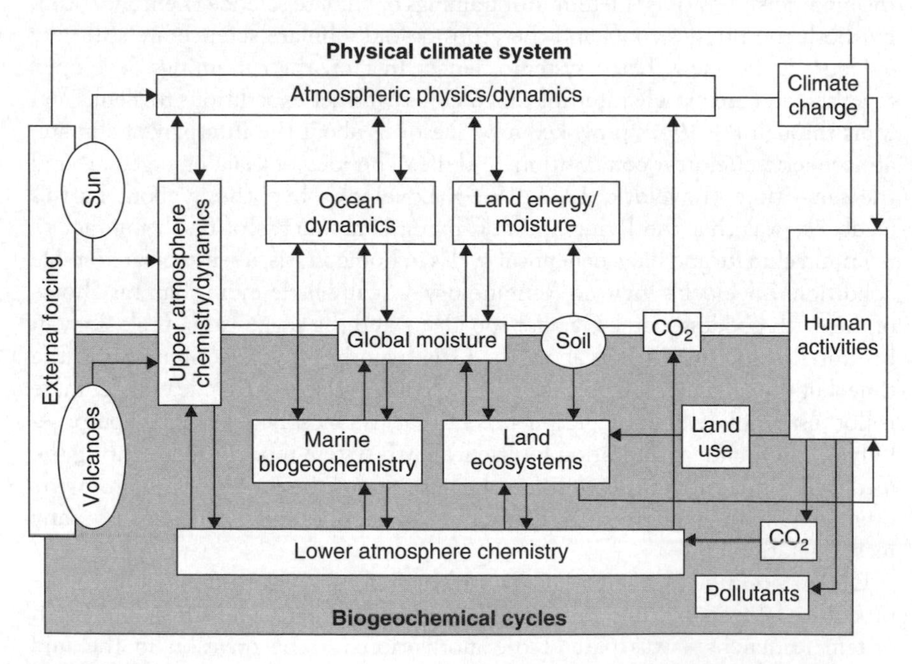

Figure 7.2 The Bretherton diagram, showing conceptual links between Earth system processes (adapted from NASA, 1986).

This may not look like deep interdisciplinarity; the tools of palaeoclimate research and contemporary climate research are essentially the same: computationally demanding global models and global observational data sets that rely on expert interpretation for their creation and use. Yet, like the separate communities engaged in model development and empirical observation of the Earth system, past and present climate scientists have tended to operate in their own separate orbits. The models may use the same fundamental equations for fluid flow, but differences in their architecture and parameterizations of processes make them all but incompatible. Over the last decade, some international researchers have made concerted efforts to coordinate across these divides in a programme called the Palaleoclimate Model Intercomparison Project (Joussaume and Taylor, 2000). There are more trivial signals of the culture divide too, though. Here is one: in many subfields, palaeoclimate data is plotted onto graphs with time on the horizontal axis, with the present day on the left (at the origin) moving rightwards into the past. Contemporary climatologists (like most other people!) plot time moving 'forwards' into the future from left to right. This makes it difficult to apprehend an issue from a quick glance at the literature, and often, that is all it takes to create a knowledge barrier.

In these recent decades, research strategies and funding for climate research (e.g. Lawton, 2001; ESF, 2003; QUEST, 2004) have focused on bringing together the physical and natural sciences, involving both the empirical and theoretical scientists, with the understanding of both the contemporary world and of the past, with its insights into the long timescale cycles in the climate system. The power of integrating all this knowledge is often emphasized in terms of the promise of solutions to the world's problems. The goal is to provide more than just explanatory science about the climate system. Earth system science apparently wants to be 'predictive' in the sense of more than merely the generation of new testable hypotheses about the world.

> One of the great scientific challenges of the 21st century is to forecast the future of planet Earth. As human activities push atmospheric carbon dioxide and methane concentrations far beyond anything seen for nearly half a million years, we find ourselves, literally, in uncharted territory, performing an uncontrolled experiment with planet Earth that is terrifying in its scale and complexity.
>
> (Lawton, 2001)

> QUEST ['Quantifying and Understanding the Earth System', a UK Natural Environment Research Council Directed Programme, 2004–2010, representing a £23M research investment] aims to achieve improved qualitative and quantitative understanding of large-scale processes and interactions in the Earth System, especially the interactions among biological, physical and chemical processes in the atmosphere, ocean and land and their implications for human activities. Thus, QUEST intends to contribute to the solution of major outstanding problems in Earth System Science. QUEST will pursue this ambition by promoting integrative, interdisciplinary activities with a

strong focus on theoretical analysis, quantitative modelling, and the systematic deployment of observational and experimental data to evaluate and improve Earth System models.

(QUEST Science Plan, 2004)

The shifting rationale for interdisciplinary working

The success of the interdisciplinary efforts in climate science so far is impressive. Contemporary Earth system models now produce awe-inspiring outputs – visualizations of the high-resolution models representing cloud-scale processes, such as Japan's Earth Simulator, now look almost satellite-photo quality (see the US University Corporation for Atmospheric Research's Visualization and Enabling Technologies Section, www.vets.ucar.edu, for some examples). More importantly, the complexity of processes that Earth system models now include (namely, the interactions between all the submodels listed in Figure 7.1) means they are giving 'realistic-looking' predictions of global changes far into the future. Climate change, as for many slow-onset or cumulative environmental hazards, is to some extent predictable – we can confidently know that physical processes taking place in nature will have consequences for social and other ecological systems.

In this context, there is a strong move towards knowledge for action, and away from 'understanding climate for its own sake'. This is the prevailing emphasis in funding strategies for the physical sciences, and is a worldwide phenomenon (after all, most countries engage in the IPCC process which has mobilized a great deal of publicly funded climate research). Yet physical science is reaching its explanatory limits in the climate context. It cannot predict confidently and precisely how and where the consequences of a changing climate will manifest themselves. In response, physical scientists have sought to extend the territory of their knowledge. This language of 'conquering' disciplines is not chosen lightly – leading scholars in the human dimensions research community have protested or drawn attention to the 'tyranny of the atmosphere' (a real bias, arising from the origins of Earth System models in weather and general circulation models, which represent the atmosphere in more detailed and complete ways than the other components of the Earth System), and noted the 'arrogance of physics' (Gibson, 2003).

What integrative efforts are being made?

The origins of global scale integrated human-biophysical modelling go back at least to the early 1970s, with the World2 and World3 models used in the Club of Rome's 'Limits to Growth' analysis. Integrated assessments focus on developing quantitative understanding of global change, using computer-based simulations of its complex dynamics, with both human and natural system components included as endogenous variables (that is, not taken as fixed external boundary conditions). In practice, this means that the global climate models have to be simple versions, or 'GCM analogue' models. Full general circulation models would be too computationally demanding, and would provide too much information for plausible interpretation of the human–environment linkages. In practice too, the 'human system component' in integrated assessments generally consists of econometric tools. The economic modules in integrated global models are computable general or partial equilibrium-seeking algorithms, and are enormously data intensive using as their input national and sectoral databases. If global data sets are not available for a given phenomenon, then the models cannot include processes relating to it. Nevertheless, some integrated assessments simulate processes of huge human significance. Hughes' (1993) International Futures simulator includes a political module, partly to ensure that government spending and revenues stay in reasonable balance, but also to represent changes in social conditions, attitudes (using input from sources such as the World Values Survey), and social organization in determining global fiscal and monetary flows. In the supporting literature, this module is described as representing the evolution of democracy and the prospects for state instability or failure.[1]

Most economists would blanch at the idea of predicting a real-world economy using general equilibrium models (see Sen, 1986 for a clear discussion that still holds even after 20 years of model and information technology development), and climate modellers would also warn against trusting the outputs of the simplified models deployed in integrated assessments. In principle, integrated assessment modellers also recognize that while their models can give insights into real structures and mechanisms, they are closed-system 'experiments', not the open-system actual world:

> Global models are not meant to predict, do not include every possible aspect of the world . . . They represent mathematical assumptions about the interrelationships among global concerns such as population, industrial output, natural resources, and pollution. Global modelers investigate what might happen if policies continue along present lines, or if specific changes are instituted.
>
> (Meadows, 1985)

The World3 model developers have conspicuously taken a consistent stance with this view, avoiding 'calibration' of the model against historic data in order not to risk spurious over precision (Meadows *et al.*, 2004). World3 has been criticized for this (Costanza *et al.*, 2007); other integrated assessment modellers are much less hesitant in assuming that the structures discovered in model runs are operating in the open system of the world. Even in the simplest integrated assessment models, it is hard to define the extent to which they are mechanistic or empirical; climate models, based on the laws of physics, tend to lean more towards the former category, and economic models more towards the latter, requiring extensive (historic) input data for the estimation of their parameters – but what are these hybrids? How real are their outputs? Despite the known and fully debated flaws in the bolt-on coupling of climate, carbon cycle and economic models, integrated assessment model outputs shape emissions scenarios, and hence mitigation policy, carbon markets, and other societal structures in the actual world. The adaptation and mitigation working groups of the Intergovernmental Panel on Climate Change (Working Groups II and III, respectively[2]) bring this information together with the explicit remit of informing policy. The way in which science does influence policy may be convoluted, but projections of climate and its consequences are not pure intellectual abstractions. The critical realist concepts of concrete conjunctures may be useful in disentangling the model world and the real world; I return to this question later.

Other integration efforts focus more on regional or sectoral consequences of climate change. This brings the issues of scale, 'nestedness' and emergence to the fore. A globally aggregated scale may be right for an analysis of the impacts of carbon dioxide emissions on global climate, but a global average that hides the presence of too much water in one locality and too little in another, or enhanced forest growth in one region and habitat losses and species extinctions in another, is meaningless (and useless, if we turn from knowledge to action). Both global and local-scale dynamics are in operation in the Earth system, and interactions and feedbacks link them; the modelling community has to grapple with the (oxymoronic?) task of simplifying the complexity, capturing those different dynamics in a tractable, comprehensible system so that the linkages between the structures and processes can be investigated (Riebsame, 1985; Costanza *et al.*, 2007).

The related domain of global ecosystem degradation shows how monetary valuation is being deployed as an integrator of sorts. Ecology is finding itself swallowed up in the knowingly reductionist (de Groot *et al.*, 2002), deliberately politicized (Pearce, 1991), explicitly anthropocentric (Millennium Ecosystem Assessment, 2005) new field of 'ecosystem services'. If we can only conceptualize nature in terms of the benefits it confers for human well-being, then it is entirely rational to manage and indeed design nature for human benefit, and to make cash payments to rebalance the distribution of those benefits or incentivise particular management options. Academic debates about novel ecosystems, which emerge as a result of human and climatic perturbations, are conspicuous by their near-absence despite the ubiquity of these ecosystems and a strong consensus that ideas of conservation and ecosystem management clearly need to address global changes and socio-ecological interplay (Hobbs *et al.*, 2006).

The impacts modelling community, which typically focuses on agriculture, water, disease vectors and so on, has been criticized for presumptions of determinism (for example, Demeritt, 2001), however the criticism does not seem to sting. Impacts modellers are sure of their starting point – climate *does* determine the Earth's biomes; models and observations and long time-series records of climate and of ecosystems all agree. You can see the logical progression: croplands are a biome (they photosynthesize and respire, live, die and decay, they require nutrients and water in predictable ways just like any other biome), so climate determines crop yield. In some ways, the construction of impacts modelling sets climate and society as opposing forces, which should open a fascinating set of philosophical inquiries about ontology, but seems not to have done so yet. In the vast majority of contemporary climate science literature, there is a focus on including more components and processes to produce more 'reliable' models. (Of course, there is also a parallel drive to obtain better and better resolution – as if looking closer is sure to make the picture prettier.) Riebsame (1985) called for 'good care in accounting for factors exogenous to the climate–society link', those difficult to quantify variables and non-linear relationships that fill the actual world, in order to produce better 'transfer functions' (mathematical representations of the causal relationship between two things). Of course climate *is* a determinant of crop yield, but in the actual world, crop yields are not caused (determined) by a regular and predictable sequence of previous climatic conditions. Critical realism invites us to be alert to this: 'Most events in open systems are conjunctures i.e. are to be explained as the result of a multiplicity of causes' (Bhaskar, 1978, p. 135).

A new focus is growing on non-deterministic modelling approaches that produce more realistic representations of human behaviour (e.g. Barthel *et al.*, 2008; Wainwright, 2007): agent-based modelling (where humans are modelled as 'agents' with specified rules governing their behaviour in a given situation), network theory and modelling, game theory and simulation, and scenario analysis. These recognise that the behaviour of components in Earth's coupled socio-ecological system are subject to interactions between the components and in some contexts to human choice and agency.

How are these integrative efforts being received?

I am writing as an interdisciplinary scholar operating mainly within the natural science community, so this section is prefaced with the caveat that I am more familiar with the literature and debates in that community than those in the social sciences. Nevertheless, the push for greater interdisciplinarity in climate research has so far been made primarily by physical scientists and climate modellers, and their funders (NERC, 2007; LWEC, 2008; ICSU, 2006; COSEPUP, 2004; EURAB, 2004). Calls for better engagement with physical climate modellers from within the social sciences are rather more like voices in the wilderness, although there have been some clear and challenging clarion calls in recent years (Hulme, 2008; Liverman and Roman Cuesta, 2008; Demeritt, 2009). Unfortunately, the physical scientists' clamouring for 'better engagement with the social sciences' often carries an implied caricature of social scientists' (presumed) skills as public communicators, and their expertise in the machinations of social engineering and political mobilization. It is not surprising that the social science community engaged in human dimensions research sometimes responds less than positively. Liverman and Roman Cuesta (2008) bring a refreshingly frank and reflexive (and practical) perspective in their thoughts on *how* to interact in interdisciplinary ways:

> For interactions between the social and earth sciences to succeed, a certain level of tolerance and mutual understanding will be needed so that social

scientists understand the earth science aspiration for quantitative socio-economic scenarios and predictions, and earth scientists understand the variations in how social scientists explain human behavior and institutions and accept the clear limits to predicting human activities and decisions.

We might cavil that the limits to prediction of human activities are not completely clear, but I want to draw attention to the fact that this entreaty hinges on the disambiguation of ontology and epistemology, a key theme in critical realism.

Some sensitivities about interdisciplinarity and 'knowledge integration' are clearly tied to power relations between the different knowledge communities. Demeritt (2001, 2009) and Hulme (2008) draw this aspect out, using the territorial tensions in the split discipline of geography as their case study. With careers on opposite sides of the human/physical geography divide, their shared insights into the perceived hierarchies of knowledge (and power) are important. The concept of a necessarily laminated system articulated and explored in critical realism might offer some comfort or sanctuary in these maneuverings.

There is a further point to probe in the way that integrative interdisciplinarity is being debated in the climate context. Physical scientists appear to hold a belief in the unity of knowledge – they seek information (data, parameters, transfer functions) about 'human dimensions' to complete their understanding of the real world. The search for understanding of human dimensions as part of a (single) causal explanation is rather different from a goal for social inquiry of plural or conjunctive explications and interpretations. Where did that 'human dimensions' phrase originate? If search engines can be trusted, Miller (1989) was the first person to apply the term and explore its scope in the context of global change. (Returning briefly to the power issue, even back then, she viewed the human component as being 'relegated' to the distinct spheres of social science and policy.) The phrase was adopted by the International Human Dimensions Programme on Global Environmental Change, that in some ways was seeking to herald the emergence of a new discipline, itself integrated or at least integratable. As mentioned in the opening section, this integration of knowledge and disciplines pertaining to the Earth system is still an active endeavour (Reid *et al.*, 2009), and this in itself creates a dynamic tension between the existing knowledge communities. Van den Besselaar and Heimeriks (2001) have studied the development of interdisciplinary fields, finding that they become discipline-like themselves over time, with their own internal connectivities, culture, and so on, even to the extent of demarcating their own boundaries and hierarchical positions. (In the climate context, we see this as vulnerability, adaptation and resilience scholars are beginning to regard themselves as operating in distinct disciplines – where does that leave anyone who wants socio-ecologically informed insights into how to navigate through any future climate crisis?) Stokols *et al.* (2008) highlight not just the integration, but the *interactivity* of disciplines in creating interdisciplinarity. Klein (1990) sees interdisciplinarity as the 'restructuring of bridges', implying that some transformative process takes place in that interaction. Strathern (2006) focuses on language in this interdisciplinary process; in the context of the deep interdisciplinarity needed for climate research, the language barrier can be all but impenetrable, with acronyms and Greek symbols on one side and polysyllabic, multi-clausal prose on the other. Strathern warns that the 'promise of a pidgin, an epistemic transfer, affecting the very knowledge base on which one works' may suppress the constructive and precise debates that ultimately determine the progress of interdisciplinarity and that are essential for any transformation.

What are the challenges? And what can CR bring?

Interdisciplinary climate research is frontier research

Whether or not interdisciplinary climate research offers a real 'solution' to the 'climate problem' – and I should be open about my belief that if solutions exist, they will require bridging or transformation of today's disciplines – it is undeniable that it offers novel directions and is innovating new research approaches. New need not mean threatening; Nissani (1997) gives us a very enjoyable pep talk, highlighting the delights and opportunities in interdisciplinary working. Still, these new directions and approaches require some care and consideration.

Critical realism offers some sturdy handholds for this less-trodden pathway. First is its provision of the concepts and discipline (in the sense of a habit or method of conduct, not a branch of knowledge) for the disambiguation of ontology and epistemology. Even if trying to understand the interlinked socio-ecological system within which we exist does not entangle these, the truth is that most physical scientists are not habituated to reflection and a thoughtful consideration of their worldviews. They are trained in the idea that science is objective and value-free. That is to say, epistemic values are acceptable because they guide scientific research itself (and training and habit mean they do not have to be thought about explicitly), but the idea of cultural values borders on taboo, so many physical scientists enter interdisciplinary areas unequipped for critical reflection on the knowledge creation or transformation process. I do not want to be alarmist about the emergence of Frankensteinian disciplines – after all, experimenting with cross-fertilization of ideas at the borders of disciplines is a very effective way to push ahead at knowledge frontiers, but it bothers me that a research community can bolt conceptual tools together because they use the same data format or grid-scale or statistical approximation without any open, deliberate consideration of what their workings really mean. Returning to a previous example, the numerical coupling of mechanistic (predictive, 'certain') models with empirical (historic, contingent, uncertain) models in some integrated assessments begs the question of what their output is, and is for. Future integrations of social sciences with physical sciences call for new conceptualizations, not an uncritical mishmash of factoids and method recipes, nor just a 'downstream' social translation of 'upstream' scientific facts. Interdisciplinarians cannot absolve themselves of the responsibility for thinking about what they are doing.

Systems theory, with its promise for elucidation of the interconnectedness of everything at all scales (Clifford and Richards, 2005) is one such 'new' conceptualization which has proliferated widely over the fifty years since its articulation, yet there is something of a rhetoric/reality gap in whole-systems research. Downy (2009, p. 3), focusing on the widespread adoption of systems jargon and overarching concepts in both the social and physical sciences, hints that disciplines may be divided by their common language:

> Both systems and complexity theories have become popular with interdisciplinary and integrated researchers, including in Earth system science, who bemoan the over-specialised reductionist view and see them as an approach to bridging the 'cultural divide'. . . Von Bertalanffy saw this too and describes systems theory as a 'broad view' which he says can be extended to the physical and natural sciences as well as the arts and humanities. Perhaps it is a good mechanism but due to its broad applicability it hasn't yet found a home among the disciplines; consequently knowledge is not consolidated but is developed disparately, and perhaps repetitively.

I have already mentioned the implicit presumptions in many climate research arenas of unity of knowledge, notably in modelling. This may be a further brake on the development of systemic understanding of our Earth. Evans and Marvin (2004, p. 24) were not talking about climate science, but the issues in sustainable development are closely related, and their concerns echo those of many human dimensions researchers in the global change domain:

> The assumption is that there is just one problem and that, by approaching it from many different sides, we can build up a complete picture that will enable an accurate and effective policy response to be developed. At this point, those who are sympathetic to the cause of feminism, post-colonialism and other moves towards standpoint epistemologies will feel the deadly chill of the grand narrative.

Here, the handhold that critical realism offers is its recognition of a laminated reality. Bhaskar (2008) clearly relishes the idea that 'in a multi-determined, multi-levelled, multi-linear, multi-relational, multi-angular, multi-perspectival, multiply determined and open pluriverse, emergence situates the widespread phenomena of dual, multiple, complex and open control'. Critical realism may offer a philosophical space for social and natural scientists (and their fruitful hybrids, and maybe even their alarming chimerae) to explore apparent dichotomies and the current artificially reinforced dialectic.[3]

Models and reality

Climate modellers 'solve equations' but they also 'simulate climate'. Their work is mostly underpinned by a positivist or empiricist philosophy, and their science is to some extent about discovering fundamental laws and increasing predictive power. But at the same time, climate change, and the conceptualization of Earth as a complex, co-adaptive system of interlinked physical, ecological and human components demands in-depth explanation, understanding and interpretation. A critical realist climate modeller would therefore emphasize methodological pluralism, and seek to explain how things are and also *why* they are as they are.

However, the enormous past investment and weighty infrastructure embedded in global climate models means that methodological pluralism is not easy. New conceptualizations suggest that new models should be created, but because of this heavy legacy, in climate modelling more than in many disciplines the push for integration tends to be directed towards creating new boxes and arrows for old models. Some people argue that climate models are themselves interdisciplinary tools, a leveller of the playing field – whatever your insight, whatever the process you want to investigate, just translate it into Fortran and you will be able and welcome to join in the game. Even when new models are created, they are designed so that their output climatologies match those of their progenitors (e.g. Pope, 2006). After all, given the comparatively short period of existence of climate modelling and Earth observation, a continuity of understanding is needed. We never construct our knowledge from scratch: it is a 'given product, a social transmit' (Bhaskar, 1978, p. 148), but we should be able to tweak or mould it into new configurations, or draw from among the ensemble of ideas that make it up, or discard concepts and tools that are not quite fit for purpose. Knowledge can be changed – but changing climate models is a difficult proposition.

The question of how *real* climate simulations are can also be a taxing one. Observational data are used to calibrate or tune the models, and also to validate them (check that their output is a plausible representation of reality). Good practice is to validate models with independent data sets from those used for calibration, but global observations come from global observing systems. Just as the magnitude of past investment locks us into the continued use of existing global climate models, the literally astronomical costs of creating Earth observation systems means that there is a strong impetus to consolidate them into a global network or partnership. The benefit is that data coverage is adequate for the task of data/model comparisons; the cost is that calibration and validation data sets are often the same. This is part of the reason for the push for reconstructions of past climates from ice cores, ocean sediments, pollen records and so on. Palaeoclimate data may provide an independent constraint on the parameters of the contemporary climate system. Debates focus very much on epistemological challenges (presumptions of uniformitarianism, the pitfalls of even sparser (and more interpreted) data coverage of the past than of the present), leaving aside ontological aspects of model/data comparison and the extension of interpretations of findings into the real world.

When climate is reconstructed, for instance from palaeodata or by regional downscaling from global models, it still needs to be situated within its cultural context. Wilbanks and Kates (1999) note the mismatch between global structure and local agency. Unfortunately one of their main suggestions for responding to this is to formalize a protocol for local studies of global climate change to increase the comparability and 'additivity' of the knowledge produced. They are still looking for empirical or positivist laws, in a context where a deeper explication is essential. Critical realism's four-planar account of social being would help bridge their gap, by bringing a focus on the personal and social relations and the subjectivity of the agent as well as on the material transactions with nature.

This brings us back to the questions of the power relations in interdisciplinarity. Demeritt (2001) has probed the instrumentalist politics of reductionist climate modelling (he also addresses the political aspects of the scaling issues in regional impact assessment modelling). In part because of the cumulative investment in climate models, and in part because of the history of science/policy interactions, 'there is no alternative' (the TINA problem) to models as the primary means for understanding global climate change. Given this, the issues of uncertainty in climate modelling are a serious concern. The climate modelling community's response is a proliferation of 'MIPs' – model intercomparison projects (PMIP is for palaeoclimate, C4MIP is for coupled carbon cycle/climate models, and so on). The tacit assumption is that good agreement in model processes in a MIP exercise tells us something about the real world. What it seems to do is focus uncomfortable attention on outlier models, and trigger debates about a climate modelling meritocracy, in which only 'good' models (those that agree with each other) are used for real-world policy engagement.

Policy – action orientation?

Science informs values, which direct actions that motivate science. Bhaskar has articulated this in terms of fact-value/theory–practice helices (Bhaskar, 1986). In the mainstream climate research community, scientists generally choose not to admit the value impregnation of factual discourse. I admit this is a statement of a strongly scientistic stance, but the tagline of the Intergovernmental Panel on Climate Change, 'policy relevant but not policy prescriptive', is taken as a mantra by thousands of scientists involved in the collective endeavour of climate research synthesis. (Not all scientists necessarily take it as a personal guiding principle, of course, but they do endorse the idea enough to be involved, if they think of it at all.) There are international differences in this stance. The power of North American lobbyists, for example, has resulted in a situation where 'advocacy' is a dirty word for scientists, whereas the UK government's Chief Scientific Advisor has given guidance to government departments that a wide enough range of views should be sought to ensure that the evidence base for policy is robust: UK scientists are strongly encouraged to engage fully. Even in the UK, though, scientists stick to the view of themselves as neutral, dispassionate, value-free observers of the system. However, scientists are often keener on the idea that facts should impregnate (other people's) values . . . Climate science literature is peppered with the idea that better knowledge (from more intricate and higher resolution models, or a better global network of trace gas observation sites) will lead to right action.[4] With its emphasis on the causal effects of social structures, can critical realist approaches in the social sciences help to provide ways forward for resolving the knowledge/action dichotomy with respect to global warming?

'All human action depends upon our ability to identify causes in open systems' (Bhaskar, 1978, p. 117). Climate research gives us foresight into future developments. How is this 'reality'? Critical realism does not leave this question hanging. Ontologically, it can accommodate the existence of structures and mechanisms, *and* their causal powers and tendencies. Critical realism extends to the postulation of what would be needed in order to cause a phenomenon to be generated. These structures and events may not yet be observable empirically, but they will be. Given the cumulative nature of the climate problem, and the long time lags, global-scale feedbacks and teleconnections in the system, the reality of a changing climate would be unknowable without climate science in its present configuration. Schneider (2005) points out the presence of both high certainty and deep uncertainty in climate understanding, and of course climate is deeply political. Uncertainty itself is an umbrella term that really does need to be unpicked and deliberated in each new context. Some might think these characteristics demand a philosophical, ethical consideration, but these dialogues are still in their infancy, and very often peripheral or detached from the science policy arena. Critical realism, in the account of the transitive and intransitive dimensions of science, states clearly that recognition of the contingency of historical and cultural contexts in limiting scientific work can logically be combined with an assertion of the reality of the objects of study and the rational evaluation of the adequacy of competing explanations, theories and models. In the current cultural context whereby grave concern about climate change is driving an agenda for more interdisciplinary research, contributions in this volume have indicated the potential of critical realism for helping us understand what is at stake in interdisciplinary science and how we might make more informed choices in managing interdisciplinary research.

Notes

1 www.ifs.du.edu/introduction/models/socio-political.aspx
2 Impacts, Adaptation, and Vulnerability, www.ipcc-wg2.org, and Mitigation of Climate Change www.ipcc-wg3.de.
3 The Intergovernmental Panel on Climate Change has divided the overall climate issue into three study areas: the 'scientific basis', impacts (now also including vulnerability and adaptation), and responses (technology, mitigation). This clearly reflected disciplinary boundaries – a look at the contents table of the synthesis reports shows this starkly, with physics, ecology, socio-political sciences and economics in neat sequence. It also reinforces and perpetuates the cultural divides in climate research.
4 See the WWF-UK report 'Weathercocks, etc.' for a critical analysis of this view.

References

Annan, J., Hargreaves, J., Lunt, D., Ridgwell, A., Rutt, I. and Sander, R. (2008). The new EGU journal: Geoscientific Model Development. GMD White Paper, Copernicus

GmbH/European Geophysical Union. Available from: www.paleo.bris.ac.uk/~ggdjl/GMD/GMD.pdf. Accessed 2 July 2009.

Barthel, R., Janisch, S., Schwarz, N., Trifkivic, A., Nickel, D., Schulz, C. and Mauser, W. (2008). An integrated modelling framework for simulating regional-scale actor responses to global change in the water domain. *Environmental Modelling and Software*, **23**(9), 1095–1121.

Bhaskar, R. (1978). *A Realist Theory of Science*. 2nd edn. Brighton, UK: The Harvester Press.

Bhaskar, R. (1986). *Scientific Realism and Human Emancipation*. London: Verso, Chapter 2.

Bhaskar, R. (2008). *Dialectic: The Pulse of Feedom*. Abingdon, UK: Routledge.

Boyle, R. (1673). *Tracts Consisting of Observations about the Saltness of the Sea: An Account of a Statical Hygroscope and Its Uses: Together with an Appendix about the Force of the Air's Moisture*. Royal Society, London.

Clifford, N. and Richards, K. (2005). Earth system science: an oxymoron? *Earth Surface Processes and Landforms*, **30**, 379–383.

COSEPUP (2004). Facilitating interdisciplinary research. National Academies Committee on Science, Engineering, and Public Policy, National Academies Press, Washington DC, USA. ISBN 13: 978 0 309 09435 1. Read online (openbook) www.nap.edu/catalog.php?record_id=11153#toc.

Costanza, R., Leemans, R., Boumans, R. and Gaddis, E. (2007). Integrated global models. In Costanza, R., Graumlich, L. and Steffen, W. (eds.). *Sustainability or Collapse? An Integrated History and Future of People on Earth*, pp. 417–446. Cambridge, MA: MIT Press.

De Groot, R.S., Wilson, M.A. and Boumans, R.M.J. (2002). A typology for the classification, description and valuation of ecosystem functions, goods and services. *Ecological Economics*, **41**, 393–408.

Demeritt, D. (2001). The construction of global warming and the politics of science. *Annals of the Association of American Geographers*, **91**, 307–337.

Demeritt, D. (2009). Geography and the promise of integrative environmental research. *Geoforum*, **40**(2), 127–129.

Downy, C.J. (2009). The academic divide between 'humans and nature' in the context of Earth System Science. QUEST Working Paper, July 2009. University of Bristol, UK.

ESF (2003). Global change: global problems, global science – Europe's contribution to global change research. European Science Foundation, *ESF Forward Look* 1, May 2003. www.esf.org/nc/publications/forward-looks.html (accessed 30 July 2009).

EU Research Advisory Board (2004). Interdisciplinarity in research. europa.eu.int/comm/research/eurab/index_en.html.

Evans, R. and Marvin, S. (2004). Disciplining the sustainable city: moving beyond science, technology or society? Paper presented at the Leverhulme International Symposium on the Resurgent City, London School of Economics, April 2004. Available from: www.lse.ac.uk/collections/resurgentCity/Papers/marvinevans.pdf. Accessed 4 July 2009.

Gibson, J.M. (2003). Arrogance – a dangerous weapon of the physics trade. *Physics Today*, **56** (2), 54–55.

Hobbs, R.J. *et al.* (2006). Novel ecosystems: theoretical and management aspects of the new ecological world order. *Global Ecology and Biogeography*, **15**, 1–7.

Hughes, B. (1993). *International Futures*, 1st edn. Boulder: Westview Press. (Now in 3rd edn, 1999.)

Hulme, M. (2008). Geographical work at the boundaries of climate change. *Transactions of the Institute of British Geography*, NS 33 5–11.

ICSU (2006). Strengthening international science for the benefit of society. ICSU Strategic Plan 2006–11. Paris. Available from www.icsu.org. Accessed 1 October 2008.

IPCC (1990, 1992). Climate change: the IPCC 1990 and 1992 assessments (combined summary publication). Intergovernmental Panel on Climate Change, Geneva, Switzerland, p. 168.

Joussaume, S. and Taylor, K.E. (2000) The Paleoclimate Modeling Intercomparison Project. Proceedings of the third PMIP workshop, Canada, 4–8 October 1999, in WCRP-111, WMO/TD-1007, edited by P. Braconnot, 9–24. http://pmip.lsce.ipsl.fr/.

Klein, J.T. (1990). *Interdisciplinarity: History, Theory and Practice*. Detroit, USA: Wayne State University Press.

Lawton, J. (2001). Earth system science. *Science*, **292**(5524), 1965.

Liverman, D.M. and Roman Cuesta, R.M. (2008). Human interactions with the Earth system: people and pixels revisited. *Earth Surface Processes and Landforms*, **33**, 1458–1471.

Lovelock, J.E. (1972). Gaia as seen through the atmosphere. *Atmosphere Environment*, **6**(8), 579–580.

Lovelock, J.E. (1979) *Gaia: A New Look at Life on Earth*. Oxford: Oxford University Press.

LWEC (2008). Living with environmental change: programme brochure, version 4. Available for download from www.nerc.ac.uk/research/programmes/lwec/resources.asp (accessed 1 August 2009).

Manabe, S. and Wetherald R. (1975). Effects of doubling CO_2 concentration on climate of a general circulation model. *Journal of the Atmospheric Sciences*, **32**(1), 3–15.

Meadows, D.H. (1985). Charting the way the world works. *Technology Review*, **88**, 55–56.

Meadows, D.H., Randers, J. and Meadows, D.L. (2004). *Limits to Growth: The 30-year Update*. Post Mills, VT: Chelsea Green.

Miller, R.B. (1989). Human dimensions of global environmental change. In Defries, R.S. and Malone, T.F. (eds.) *Global Change and Our Common Future*. Papers from a Forum Meeting, Washington DC, USA, May 2–3, 1989. National Academy Press: Washington DC, USA. p84–89 Available on www.nap.edu/openbook.php?record_id =1411&page=84.

NASA (1986). Earth system science overview. Earth System Sciences Committee of the National Aeronautics and Space Administration Advisory Council (F. Bretherton *et al.*), Washington DC, USA.

NERC (2007). Next generation science for Planet Earth. www.nerc.ac.uk/publications/strategicplan/nextgeneration.asp.

Nissani, M. (1997). Ten cheers for interdisciplinarity: the case for interdisciplinary knowledge and research. *The Social Science Journal*, **34**(2), 201–216.

Oreskes, N., Shrader-Frechette, K. and Belitz, K. (1994). Verification, validation, and confirmation of numerical models in the earth sciences. *Science*, **263**, 641–646.

Pannabecker, J.R. (1991). Technological impacts and determinism in technology education: alternate metaphors from social constructivism. *Journal of Technology Education*, **3**(1). Downloadable from scholar.lib.vt.edu/ejournals/JTE/v3n1/html/pannabecker.html.

Pope, V. (2006). The Hadley Centre climate model HadGEM1. www.bom.gov.au/bmrc/basic/wksp18/papers/Pope_HadGEM1.pdf.

QUEST (2004). Quantifying and understanding the Earth system: science plan. Available from the UK Natural Environment Research Council's British Atmospheric Data Centre: badc.nerc.ac.uk/data/quest/QUEST_scienceplan.pdf. Accessed 1 September 2009.

Reid, R.V., Bréchignac, C. and Yuan Tse Lee (2009). Earth system research priorities. *Science*, **325**(5938), 245.

Riebsame, W.E. (1985). Research in climate–society interaction. In Kates, R.W., Ausubel, J. and Berberian, M. (eds.) *Climate Impact Assessment — Scope 27*, Chapter 3. Chichester, UK: John Wiley & Sons.

Schneider, S. (2005). An overview of the climate change problem. stephenschneider.stanford.edu/Climate. Accessed 1 August 2009.

Sen, A.K. (1986). Prediction and economic theory. *Proceedings of the Royal Society. Mathematical, Physical and Engineering Sciences*, **407**(1832), 3–23.

Stokols, D., Hall, K.L., Taylor, B.K. and Moser, R.P. (2008). The science of team science: overview of the field and introduction to the supplement. *American Journal of Preventive Medicine*, **35**(2S), 77–89.

Strathern, M. (2006). A community of critics? Thoughts on new knowledge. *Journal of the Royal Anthropological Institute (N.S.)*, **12**, 191–209.

van den Besselaar, P.A.A. and Heimeriks, G. (2001). Disciplinary, multidisciplinary, interdisciplinary: concepts and indicators. In Davis, M. and Wilson, C.S. (eds.), ISSI 2001, 8th International Conference of the Society for Scientometrics and Informetrics, Sydney: UNSW 2001, pp. 705–716. Downloadable from hcs.science.uva.nl/usr/peter/publications/2002issi.pdf.

Wainwright, J. (2007). Can modelling enable us to understand the role of humans in landscape evolution? *Geoforum*, **39**(2), 659–674.

Wilbanks T.J. and Kates R.W. (1999). Global change in local places: how scale matters. *Climatic Change*, **433**, 601–628. Blackwell Publishing, Ltd.

8 The need for a transdisciplinary understanding of development in a hot and crowded world

Robert Costanza

The need for a transdisciplinary vision

Practical problem solving in complex, human dominated ecosystems requires the integration of three elements: (1) active and ongoing envisioning of both how the world works and how we would like the world to be; (2) systematic analysis appropriate to and consistent with the vision; and (3) implementation appropriate to the vision. Scientists generally focus on only the second of these steps, but integrating all three is essential to both good science and effective management. 'Subjective' values enter in the 'vision' element, both in terms of the formation of broad social goals and in the creation of a 'pre-analytic vision' which necessarily precedes any form of analysis. Because of this need for vision, completely 'objective' analysis is impossible. In the words of Joseph Schumpeter (1954, p. 41):

> In practice we all start our own research from the work of our predecessors, that is, we hardly ever start from scratch. But suppose we did start from scratch, what are the steps we should have to take? Obviously, in order to be able to posit to ourselves any problems at all, we should first have to visualize a distinct set of coherent phenomena as a worthwhile object of our analytic effort. In other words, analytic effort is of necessity preceded by a preanalytic cognitive act that supplies the raw material for the analytic effort. In this book, this preanalytic cognitive act will be called Vision. It is interesting to note that vision of this kind not only must precede historically the emergence of analytic effort in any field, but also may reenter the history of every established science each time somebody teaches us to see things in a light of which the source is not to be found in the facts, methods, and results of the preexisting state of the science.

The pre-analytic vision of science is changing from the 'logical positivist' view (which holds that science can discover ultimate 'truth' by falsification of hypothesis) to a more pragmatic view that recognizes that we do not have access to any ultimate, universal truths, but only to useful abstract representations (models) of the world. Science, in both the logical positivist and in this new 'pragmatic modeling' or 'critical realist' vision, works by building models and testing them. But the new vision recognizes that the tests are rarely, if ever, conclusive, the models can only apply to a limited part of the real world; and the ultimate goal is therefore not 'truth' but quality and utility. In the words of William Deming 'All models are wrong, but some models are useful' (McCoy 1994).

Both our 'pre-analytic vision' of how the human economy relates to the rest of nature and the economy itself are also changing. The human economy has passed from an 'empty world' era in which human-made capital was the limiting factor in economic development to the current 'full world' era in which remaining natural and social capital have become the limiting factors (Costanza *et al.*, 1997a). This implies a very different vision of the economy and its place in the overall system. In an empty world we could afford to ignore the interconnections and focus only on the economy, in disciplinary isolation from the rest of the system. In a full world we must recognize the interconnections and pursue a transdisciplinary synthesis. This has huge implications for how we deal with climate change.

Figure 8.1a shows the conventional, disciplinary, 'empty world' economic pre-analytic vision. The primary factors of production (land, labor, and capital) combine in the economic process to produce goods and services (usually measured as Gross Domestic Product or GDP). GDP is divided into consumption (which is the sole contributor to individual utility and welfare) and investment (which goes into maintaining and increasing the capital stocks). Preferences are fixed. In this model the primary factors are perfect substitutes for each other so 'land' (including climate and other ecosystem services) can be almost ignored. Property rights are usually simplified to either private or public and their distribution is usually taken as fixed and given. There is nothing in this model that recognizes climate as an important component of the system.

Figure 8.1b shows an alternative 'full world' view of the system (Ekins, 1992; Costanza *et al.*, 1997a). Notice that the key elements of the conventional view are still present, but more has been added and some priorities have changed. There is limited substitutability between the basic forms of capital in this model and their number has expanded to four. Their names have also changed to better reflect their roles: (1) natural capital (formerly land) includes climate, ecological systems, mineral deposits and other aspects of the natural world; (2) human capital (formerly labor) includes both the physical labor of humans and the know-how stored in their brains; (3) manufactured capital includes all the machines and other infrastructure of the human economy; and (4) social (or cultural) capital. Social capital includes the web of interpersonal connections, institutional arrangements, rules, and norms that allow individual human interactions to occur (Berkes and Folke, 1994). Property rights regimes in this model are complex and flexible, spanning the range from individual to common to public property. Natural capital captures solar energy and behaves as an autonomous complex system and the model conforms to the basic laws of thermodynamics. Natural capital contributes to the production of marketed economic goods and services, which affect human welfare. It also produces ecological services and amenities

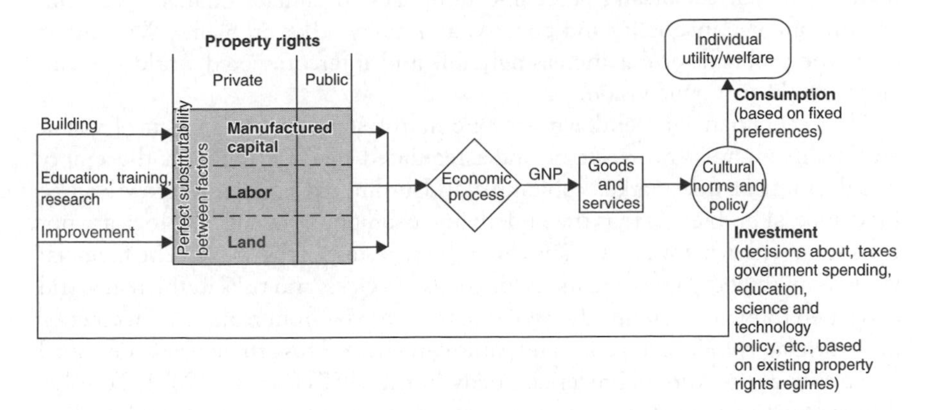

Figure 8.1a 'Empty world' economy

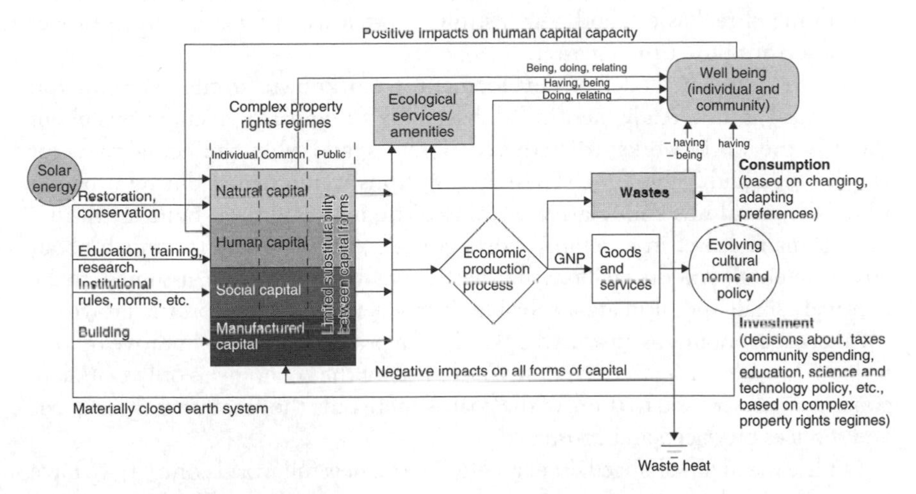

Figure 8.1b 'Full world' economy

that directly contribute to human welfare without ever passing through markets. There is also waste production by the economic process, which contributes negatively to human welfare and has a negative impact on capital and ecological services. Preferences are adapting and changing but basic human needs are constant. Human welfare is a function of much more than the consumption of conventional economic goods and services (Costanza *et al.*, 2008).

These visions of the world are significantly different. As Ekins (1992) points out: 'It must be stressed that that the complexities and feedbacks of model 2 are not simply glosses on model 1's simpler portrayal of reality. They fundamentally alter the perceived nature of that reality and in ignoring them conventional analysis produces serious errors . . .' (p. 151). These errors have now become manifest in our social and economic responses to climate change, peak oil, growing income inequality and poverty, and many other problems. We cannot solve the problems of our increasingly full and interconnected world with an empty world economic vision.

The 2008 financial meltdown was one manifestation of this misfit of vision and reality. It was the result of under-regulated markets built on the empty world vision of free market capitalism and unlimited economic growth. The fundamental problem is that the underlying assumptions of this ideology are not consistent with what we now know about the real state of the world. The financial world is, in essence, a set of markers for goods, services, and risks in the real world and when those markers are allowed to deviate too far from reality, 'adjustments' must ultimately follow and crisis and panic can ensue. This problem was identified as far back as the work of Frederick Soddy in the 1930s (Soddy, 1933). To solve this and future financial crises requires that we reconnect the markers with reality. What are our real assets and how valuable are they? To do this requires both a new vision of what the economy is and what it is for, proper and comprehensive accounting of real assets, and new institutions that use the market in its proper role of servant rather than master.

The mainstream model of development (also known as the 'Washington consensus') is, as already mentioned, based on a number of assumptions about the way the world works, what the economy is, and what the economy is for (Figure 8.1a and Table 8.1). These assumptions were created during a period when the world was still relatively empty of humans and their built infrastructure. It made sense, in an empty world context, not to worry too much about environmental and social 'externalities' since they could be assumed to be relatively small and ultimately solvable. It made sense to focus on the growth of the market economy, as measured by GDP, as a primary means to improve human welfare. It made sense, in that context, to think of the economy as only marketed goods and services and to think of the goal as increasing the amount of these goods and services produced and consumed.

But the world has changed dramatically. In the new full world context, we have to reconceptualize what the economy is and what it is for. We have to first remember that the goal of the economy is to sustainably improve human well-being and quality of life. We have to remember that material consumption and GDP are merely means to that end, not ends in themselves. We have to recognize, as both ancient wisdom and new psychological research tell us, that material consumption beyond real need can actually reduce our well-being. We have to better understand what really does contribute to sustainable human well-being, and recognize the substantial contributions of natural and social capital, which are now the limiting factors to sustainable human well-being in many countries. We have to be able to distinguish between real poverty in terms of low quality of life, and merely low monetary income. Ultimately, we have to create a new vision of what the economy is and what it is for, and a new model of development that acknowledges this new full world context and vision (Table 8.1).

Table 8.1 Basic characteristics of the current development model and the emerging sustainable and desirable 'ecological economics' development model

	Current development model: *the 'Washington Consensus'*	***Sustainable and desirable development model:*** *an emerging 'green consensus'*
Primary policy goal	**More:** economic growth in the conventional sense, as measured by GDP. The assumption is that growth will ultimately allow the solution of all other problems. More is always better.	**Better:** Focus must shift from merely growth to 'development' in the real sense of improvement in quality of life, recognizing that growth has negative by-products and more is not always better.
Primary measure of progress	Gross Domestic Product (GDP).	General Progress Indicator (GPI) (or similar).
Scale/carrying capacity	Not an issue since markets are assumed to be able to overcome any resource limits via new technology and substitutes for resources are always available. Climate change will not limit GDP growth.	A primary concern as a determinant of ecological sustainability. Natural capital and ecosystem services are not infinitely substitutable and real limits exist. Climate change will place boundaries on economic expansion and a shift to quality rather than quantity is necessary.
Distribution/ poverty	Lip service, but relegated to 'politics' and a 'trickle down' policy: a rising tide lifts all boats. Climate change not relevant.	A primary concern since it directly affects quality of life and social capital and in some very real senses is often exacerbated by growth: a too rapidly rising tide only lifts yachts, while swamping small boats. Problem is exacerbated by climate change, which will affect the poor more severely. Policies needed to link dealing with climate change and poverty reduction, like the Earth Atmospheric Trust.
Economic efficiency/ allocation	The primary concern, but generally including only marketed goods and services (GDP) and institutions, excluding social and natural capital (e.g. climate).	A primary concern, but including both market and non-market goods and services and effects. Emphasizes the need to incorporate the value of natural (i.e. climate) and social capital to achieve true allocative efficiency.
Property rights	Emphasis on private property and conventional markets. Attempts to privatize elements of the climate system.	Emphasis on a balance of property rights regimes appropriate to the nature and scale of the system, and a linking of rights with responsibilities. A larger role for common property institutions for managing common assets (like the atmosphere) in addition to private and state property.
Role of government	To be minimized and replaced with private and market institutions	A central role, including new functions as referee, facilitator and broker in a new suite of common asset institutions, like the Earth Atmospheric Trust
Principles of governance	*Laissez faire* market capitalism.	Lisbon principles of sustainable governance.

Quality of life, happiness, and the real economy

There is a substantial body of new research on what actually contributes to human well-being and quality of life (Costanza *et al.*, 2008). This new 'science of happiness' clearly demonstrates the limits of conventional economic income and consumption in contributing to well-being. Kasser (2003) points out, for instance, that people who focus on material consumption as a path to happiness are actually less happy and even suffer higher rates of both physical and mental illnesses than those who do not. Material consumption beyond real need is a form of psychological 'junk food' that only satisfies for the moment and ultimately leads to depression.

Easterlin (2005), has shown that well-being tends to correlate well with health, level of education, and marital status, and not very well with income beyond a certain fairly low threshold. He concludes that:

> People make decisions assuming that more income, comfort, and positional goods will make them happier, failing to recognize that hedonic adaptation and social comparison will come into play, raise their aspirations to about the same extent as their actual gains, and leave them feeling no happier than before. As a result, most individuals spend a disproportionate amount of their lives working to make money, and sacrifice family life and health, domains in which aspirations remain fairly constant as actual circumstances change, and where the attainment of one's goals has a more lasting impact on happiness. Hence, a reallocation of time in favor of family life and health would, on average, increase individual happiness.

Layard (2005) synthesizes many of these ideas and concludes that current economic policies are not improving happiness and that 'happiness should become the goal of policy, and the progress of national happiness should be measured and analyzed as closely as the growth of GNP'.

Frank (2000) also concludes that some nations would be better off – overall national well-being would be higher, that is – if we actually consumed less and spent more time with family and friends, working for our communities, maintaining our physical and mental health, and enjoying nature.

On this last point, there is substantial and growing evidence that natural systems contribute heavily to human well-being. Costanza *et al.* (1997b) estimated the annual, non-market value of the earth's ecosystem services at $33 trillion/yr, substantially larger than global GDP at the time and yet an almost certainly a conservative underestimate. The UN Millennium Ecosystem Assessment (2005) is a global compendium of the status and trends of ecosystem services and their contributions to human well-being.

So, if we want to assess the 'real' economy – all the things which contribute to real, sustainable, human well-being – as opposed to only the 'market' economy, we have to measure and include the non-marketed contributions to human well-being from nature, from family, friends and other social relationships at many scales, and from health and education. One convenient way to summarize these contributions is to group them into the four basic types of capital that are necessary to support the real, human-well-being-producing economy described earlier: built capital, human capital, social capital and natural capital.

The market economy covers mainly built capital (factories, offices, and other built infrastructure and their products) and part of human capital (spending on labor, health and education), with some limited spillover into the other two. It leaves out the contributions of natural capital and the ecosystem services it provides. Ecosystem services occur at many scales, from climate regulation at the global scale, to flood protection, soil formation, nutrient cycling, recreation, and aesthetic services at the local and regional scales. Dealing with climate change requires that we recognize the highly interconnected nature of the current system and the value of the climate system and other forms of natural capital.

Are we really making progress?

Given this definition of the real economy, are we really making progress? Is the mainstream development model really working, even in the 'developed' countries? One way to tell is through surveys of people's life satisfaction, which have been relatively flat in the US and many other developed countries since about 1975. A second approach is an aggregate measure of the real economy that has been developed as an alternative to GDP called the Index of Sustainable Economic Welfare (ISEW – Daly and Cobb, 1989) and more recently renamed the Genuine Progress Indicator (GPI – Cobb *et al.*, 1995)

Let's first take a quick look at the problems with GDP as a measure of true human well-being. GDP is not only limited – measuring only marketed economic activity or gross income — it also counts all of this activity as positive. It does not separate desirable, well-being-enhancing activity from undesirable well-being-reducing activity. For example, an oil spill increases GDP because someone has to clean it up, but it obviously detracts from society's well-being. From the perspective of GDP, more crime, more sickness, more war, more pollution, more fires, storms, and pestilence are all potentially good things, because they can increase marketed activity in the economy.

GDP also leaves out many things that *do* enhance well-being but are outside the market. For example, the unpaid work of parents caring for their own children at home doesn't show up, but if these same parents decide to work outside the home to pay for child care, GDP suddenly increases. The non-marketed work of natural capital in providing clean air and water, food, natural resources, and other ecosystem services doesn't adequately show up in GDP, either, but if those services are damaged and we have to pay to fix or replace them, then GDP suddenly increases. Finally, GDP takes no account of the distribution of income among individuals. But it is well-known that an additional $1 worth of income produces more well-being if one is poor rather than rich. It is also clear that a highly skewed income distribution has negative effects on a society's social capital.

The GPI addresses these problems by separating the positive from the negative components of marketed economic activity, adding in estimates of the value of non-marketed goods and services provided by natural, human and social capital, and adjusting for income-distribution effects. While it is by no means a perfect representation of the real well-being of nations, GPI is a much better approximation than GDP. As Amartya Sen and others have noted, it is much better to be approximately right in these measures than precisely wrong.

Comparing GDP and GPI for the US (Figure 8.2) shows that, while GDP has steadily increased since 1950, with the occasional dip or recession, GPI peaked in about 1975 and has been flat or gradually decreasing ever since. From the perspective of the real economy, as opposed to just the market economy, the US has been in recession since 1975. As already mentioned, this picture is also consistent with survey-based research on people's stated life-satisfaction. The US and several other developed countries are now in a period of what Herman Daly has called 'un-economic growth', where further growth in marketed economic activity (GDP) is actually reducing well-being on balance rather than enhancing it. In terms of the four capitals, while built capital has grown, human, social and natural capital have declined or remained constant and more than canceled out the gains in built capital.

A new sustainable, ecological model of development

A new model of development consistent with our new full world context (Table 8.1) would be based clearly on the goal of sustainable human well-being.

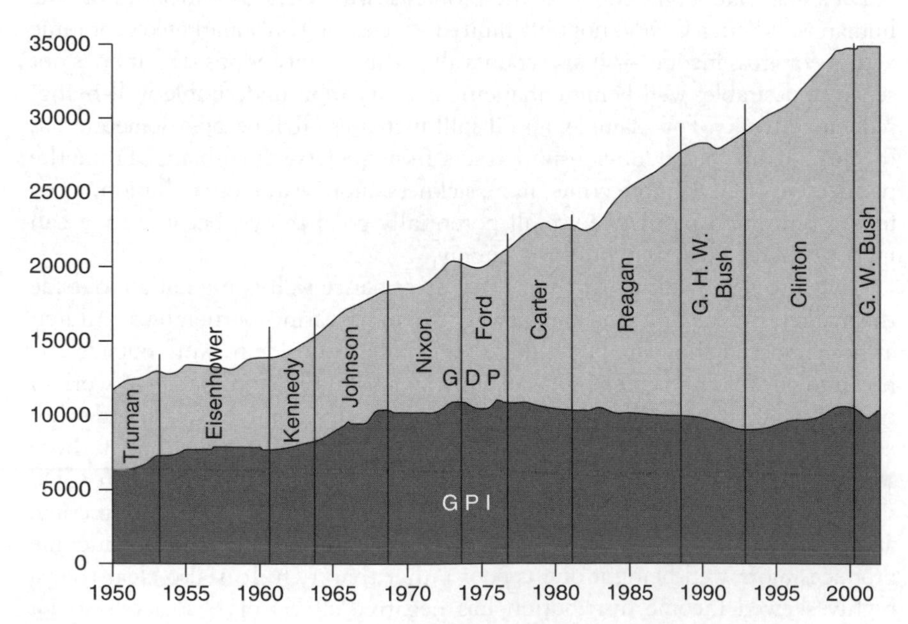

Figure 8.2 Comparing GDP and GPI for the US.

It would use measures of progress that clearly acknowledge this goal (i.e. GPI instead of GDP). It would acknowledge the importance of ecological sustainability, social fairness, and real economic efficiency.

Ecological sustainability implies recognizing that natural and social capital are not infinitely substitutable for built and human capital, and that real biophysical limits exist to the expansion of the market economy. Climate change is perhaps the most obvious and compelling of these limits.

Social fairness implies recognizing that the distribution of wealth is an important determinant of social capital and quality of life. The conventional development model, while explicitly aimed at reducing poverty, has bought into the assumption that the best way to do this is through growth in GDP. This has not proved to be the case and explicit attention to distribution issues is sorely needed. As Frank (2007) has argued, economic growth beyond a certain point sets up a 'positional arms race' that changes the consumption context and forces everyone to consume too much of easily seen positional goods (like houses and cars) at the expense of non-marketed, non-positional goods and services from natural and social capital. Increasing inequality of income actually reduces overall societal well-being, not just for the poor, but across the income spectrum.

Real economic efficiency implies including all resources that affect sustainable human well-being in the allocation system, not just marketed goods and services. Our current market allocation system excludes most non-marketed natural and social capital assets and services that are huge contributors to human well-being. The current development model ignores this and therefore does not achieve real economic efficiency. A new, sustainable ecological development model would measure and include the contributions of natural and social capital and could better approximate real economic efficiency.

The new development model would also acknowledge that a complex range of property rights regimes are necessary to adequately manage the full range of resources that contribute to human well-being. For example, most natural and social capital assets are public goods. Making them private property does not work well. On the other hand, leaving them as open access resources (with no property rights) does not work well either. What is needed is a third way to *propertize* these resources without privatizing them. Several new (and old) common property rights systems have been proposed to achieve this goal, including various forms of common property trusts.

For example, one proposed institution aimed at massively reducing global carbon emissions and at the same time reducing poverty is the 'Earth Atmospheric Trust' (Barnes *et al.*, 2008). The system would include six basic elements:

(i) A global cap-and-trade system for all greenhouse gas emissions. A cap-and trade system is preferable to a tax, because caps set quantity (the ultimate goal) and allow price to vary; taxes set price and allow quantity to vary.

(ii) Auctioning off *all* emission permits before allowing trading among permit holders. This essential feature will send the right price signals to emitters.

(iii) Reducing the cap over time to stabilize concentrations of greenhouse gases in the atmosphere at a level equivalent to 350 parts per million of carbon dioxide.
(iv) Depositing all the revenues into an Earth Atmospheric Trust, transparently administered by elected trustees serving long terms and provided with a clear mandate to protect Earth's climate system and atmosphere for the benefit of current and future generations.
(v) Returning a fraction of the revenues derived from auctioning permits to all people on Earth in the form of a per capita payment.
(vi) Use of the remainder of the revenues to enhance and restore the atmospheric asset, to invest in both social and technological innovations, to assist developing countries, and to administer the Trust.

The role of government also needs to be reinvented. In addition to government's role in regulating and policing the private market economy, it has a significant role to play in expanding the 'commons sector', that can propertize and manage non-marketed natural and social capital assets. It also has a major role to play as facilitator of societal development of a shared vision of what a sustainable and desirable future would look like. Strong democracy based on developing a shared vision is an essential prerequisite to building a sustainable and desirable future (Prugh *et al.*, 2000). This new vision implies a core set of principles for sustainable governance.

Principles of sustainable governance

The key to achieving sustainable governance in the new full world context is an integrated (across disciplines, stakeholder groups, and generations) approach based on the paradigm of 'adaptive management', whereby policy-making is an iterative experiment acknowledging uncertainty, rather than a static 'answer'. Within this paradigm, six core principles (the Lisbon principles) that embody the essential criteria for sustainable governance have been proposed (Costanza *et al.*, 1998). Some of them are already well accepted in the international community (for example, Principle 3); others are variations on well-known themes (for example, Principle 2 is an extension of the subsidiary principle); while others are relatively new in international policy, although they have been well developed elsewhere (for example, Principle 4). The six Principles together form an indivisible collection of basic guidelines governing the use of common natural and social capital assets.

- *Principle 1: Responsibility*. Access to common asset resources carries attendant responsibilities to use them in an ecologically sustainable, economically efficient, and socially fair manner. Individual and corporate responsibilities and incentives should be aligned with each other and with broad social and ecological goals.

- *Principle 2: Scale-matching*. Problems of managing natural and social capital assets are rarely confined to a single scale. Decision-making should (i) be assigned to institutional levels that maximize input, (ii) ensure the flow of information between institutional levels, (iii) take ownership and actors into account, and (iv) internalize costs and benefits. Appropriate scales of governance will be those that have the most relevant information, can respond quickly and efficiently, and are able to integrate across scale boundaries.
- *Principle 3: Precaution*. In the face of uncertainty about potentially irreversible impacts to natural and social capital assets, decisions concerning their use should err on the side of caution. The burden of proof should shift to those whose activities potentially damage natural and social capital.
- *Principle 4: Adaptive management*. Given that some level of uncertainty always exists in common asset management, decision-makers should continuously gather and integrate appropriate ecological, social and economic information with the goal of adaptive improvement.
- *Principle 5: Full cost allocation*. All of the internal and external costs and benefits, including social and ecological, of alternative decisions concerning the use of natural and social capital should be identified and allocated. When appropriate, markets should be adjusted to reflect full costs.
- *Principle 6: Participation*. All stakeholders should be engaged in the formulation and implementation of decisions concerning natural and social capital assets. Full stakeholder awareness and participation contributes to credible, accepted rules that identify and assign the corresponding responsibilities appropriately.

Some policies to achieve real, sustainable development

The conventional development model is not working, for either the developed or the developing world. It is not sustainable and it is also not desirable. It is based on a now obsolete empty world vision and it is leading us to disaster.

We need to accept that we now live in a full world context where natural and social capital are the limiting factors. We could achieve a much higher quality of life, and one that would be ecologically sustainable, socially fair, and economically efficient, if we shift to a new sustainable development paradigm that incorporates these principles.

The problem is that our entire modern global civilization is, as even former President Bush has acknowledged, 'addicted to oil' and addicted to consumption and the conventional development model in general. An addictive substance is something one has developed a dependence on, which is either not necessary or harmful to one's longer-term well-being. Fossil fuels (and excessive material consumption in general) fit the bill. We can power our economies with renewable energy, and we can be happier with lower levels of consumption, but we must first break our addiction to fossil fuels, consumption, and the conventional development model, and as any addict can tell you: 'that ain't easy'. But in order to break an addiction of any kind, one must first clearly see the benefits of breaking it, and the costs of remaining addicted, facts that accumulating studies like the IPCC reports, the Stern Review (2007), the Millennium Ecosystem Assessment (2005) and many others are making more apparent every day.

What else can we do to help break this addiction? Here are just a few suggestions.

- Create and share a vision of a future with zero fossil fuel use and a quality of life higher than today. That will involve understanding that GDP is a means to an end, not the end itself, and that in some countries today more GDP actually results in less human well-being (while in others the reverse is still true). It will require a focus on sustainable scale and just distribution. It will require an entirely new and broader vision of what the economy is, what it's for, and how it functions
- Convene a 'new Bretton Woods' conference to establish the new measures and institutions needed to replace GDP, the World Bank, the IMF, and the WTO. These new institutions would promote:
- Shifting primary national policy goals from increasing marketed economic activity (GDP) to maximizing national well-being (GPI or something similar). This would allow us to see the interconnections between built, human, social and natural capital, and build real well-being in a balanced and sustainable way.
- Reforming tax systems to send the right incentives by taxing negatives (pollution, depletion of natural capital, overconsumption) rather than positives (labor, savings, investment).
- Expanding the commons sector by developing new institutions that can *propertize* the commons without privatizing them. Examples include various forms of common asset trusts, like the atmospheric (or sky) trust (Barnes *et al.*, 2008) payments for depletion of natural and social capital and rewards for protection of these assets.
- Reforming international trade to promote well-being over mere GDP growth. This implies protecting natural capital, labor rights, and democratic self-determination first and *then* allowing trade, rather than promoting the current trade rules that ride roughshod over all other societal values and ignore non-market contributions to well-being.

Conclusions

We can break our addiction to fossil fuels, overconsumption, and the current development model and create a more sustainable and desirable future with a stable climate and shared prosperity. It will not be easy, it will require a new transdisciplinary vision, new measures, and new institutions. It will require an economics that is fully engaged and integrated with the full range of other disciplines in a true transdisciplinary synthesis. It will require a directed evolution of our entire society (Beddoe *et al.*, 2009). But it is not a sacrifice of quality of life to break this addiction. Quite the contrary, it is a huge sacrifice not to.

References

Barnes, P., Costanza, R., Hawken, P., Orr, D., Ostrom, E., Umana, A. and Young, O. (2008). Creating an earth atmospheric trust. *Science*, **319**, 724.

Beddoe, R., Costanza, R., Farley, J., Garza, E., Kent, J., Kubiszewski, I., Martinez, L., McCowen, T., Murphy, K., Myers, N., Ogden, Z., Stapleton, K. and Woodward, J. (2009). Overcoming systemic roadblocks to sustainability: the evolutionary redesign of worldviews, institutions and technologies, *Proceedings of the National Academy of Sciences*, **106**, 2483–2489.

Berkes, F. and Folke, C. (1994). Investing in cultural capital for a sustainable use of natural capital. In A.M. Jansson, M. Hammer, C. Folke and R. Costanza (eds.) *Investing in Natural Capital: the Ecological Economics Approach to Sustainability*, pp. 128–149. Washington, DC: Island Press.

Cobb, C., Halstead, T. and Rowe, J. (1995). *The Genuine Progress Indicator: Summary of Data and Methodology*, San Francisco: Redefining Progress.

Costanza, R., Cumberland, J.C., Daly, H.E., Goodland, R. and Norgaard, R. (1997a). *An Introduction to Ecological Economics*. Boca Raton: St. Lucie Press.

Costanza, R., d'Arge, de Groot, R., Farber, S., Grasso, M., Hannon, B., Naeem, S., Limburg, K., Paruelo, J., O' Neill, R.V., Raskin, R., Sutton, P. and van den Belt, M. (1997b). The value of the world's ecosystem services and natural capital. *Nature*, **387**, 253–260.

Costanza, R., Andrade, F., Antunes, P., van den Belt, M., Boersma, D., Boesch, D.F., Catarino, F., Hanna, S., Limburg, K., Low, B., Molitor, Pereira, M.G., Rayner, S., Santos, R., Wilson, R.J. and Young, M. (1998). Principles for sustainable governance of the oceans, *Science*, **281**, 198–199.

Daly, H. E. and Cobb, J. (1989). *For the Common Good*. Boston: Beacon Press.

Easterlin, R.A. (2003). Explaining happiness, *Proeedings of the Naional Academy of Sciences, USA*, **100**, 11176–11183.

Ekins, P. (1992). A four-capital model of wealth creation. In *Real-life Economics: Understanding Wealth Creation*, ed. P. Ekins and M. Max-Neef, pp. 147–155. London: Routledge.

Fisher, B., Ali, S., Beer, C., Bond, L., Boumans, R., Danigelis, N.L., Dickinson, J., Elliott, C., Farley, J., Gayer, D.E., Hudspeth, T., MacDonald, L.G., Mahoney, D., McCahill, L., McIntosh, B., Reed, B., Rizvi, S.A.T., Rizzo, D.M., Simpatico, T. and Snapp, R. (2008). An integrative approach to quality of life measurement, research, and policy. *Surveys and Perspectives Integrating Environment and Society*. **1**, 1–5.

Frank, R. (2000). *Luxury Fever*. Princeton, NJ: Princeton University Press.

Frank, R. (2007). *Falling Behind: How Rising Inequality Harms the Middle Class*. Berkeley: University of California Press.

Kasser, T. (2003). *The High Price of Materalism*. Cambridge, MA: MIT Press

Layard, R. (2005). *Happiness: Lessons From a New Science*. Penguin.

McCoy, R. (1994). *The Best of Deming*. Knoxville, TN: SPC Press.

Millennium Ecosystem Assessment (2005). *Ecosystems and Human Well-being: Synthesis*. Washington, DC: Island Press.

Prugh, T., Costanza, R. and Daly, H. (2000). *The Local Politics of Global Sustainability*. Washington, DC: Island Press.

Soddy, F. (1933). *Wealth, Virtual Wealth, and Debt: The Solution of the Economic Paradox*. Dutton.
Schumpeter, J. (1954). *History of Economic Analysis*. London: Allen & Unwin, 1260 pp.
Stern, N. (2007). *The Economics of Climate Change: The Stern Review*. Cambridge, UK: Cambridge University Press.

9 Knowledge, democracy and action in response to climate change

Kjetil Rommetveit, Silvio Funtowicz and Roger Strand

Introduction

Climate change has entered at the top of the national and international political agendas. The urge for change is global in more than one sense of the word:

- First, the problem encompasses all countries, and so seems to be demanding world-wide communication and understanding.
- Second, it cuts through literally every sector of society, posing huge problems for the co-ordination of action across a number of sectors traditionally separated.
- Third, it potentially concerns every aspect of our lives, be that as parents, professionals, community members, citizens, habitants or political subjects, not to mention our roles as consumers of goods, transportation, services and culture.

There are many problematic aspects related to these challenges, many of which relate to their complexity, interconnectedness and far-reaching implications across a number of scales and boundaries of disciplinary, institutional and national character. Many of these problems are noted and accepted within most dominant approaches to climate change policy. However, we contend in this chapter, at the same time as such problems are recognised, most policy approaches are hampered by a lack of appreciation of the cultural and democratic character of scientific knowledge and of the difficult relations of knowledge to action. Whereas many recognise these issues as central, they are more often than not left out of the analytical framework. The risk is therefore real that we, even with the best of intentions, end up reproducing or exacerbating predominant problems, or even produce new ones.

This appears even more so when we consider the mounting political pressures building up as the sense of climate urgency increases. The growing sense of urgency is mainly brought forward through the increasing acceptance and attention given to the IPCC reports. But that attention can hardly be separated from increasing numbers of reports about natural catastrophes and extreme weather from all over the world, such as Hurricane Katrina, the 2004 tsunami, floods in Latin America and Europe, drought and desertification in Africa and the Middle East. Even less so can these events be separated from the attention given them in the media: nationally and globally, climate change has taken centre stage in the headlines. Therefore, even if we do not *know* whether hurricane Katrina was caused by global warming, that particular association is becoming increasingly easy to make. As noted by Mario Giampetro (2008), the granting of the Nobel Peace Prize and the Oscar to one and the same person says something about the unique position and context of the climate change issue.

Within this situation there is a growing danger of *policy vacuums*, i.e. a growing sense of urgency coupled with a lack of knowledge of what to do and a lack of institutions where the issues could be addressed. The mere feeling of environmental crisis may pave the way for states of exception (Agamben, 2005), and open up the scope for Machiavellian politics (Beck, 1998) in which power goes unchecked. The climate change problem is a highly politicised problem, the potential of which is amplified manifold through its intimate connection with the powerful sectors of energy, industry and technology.

The main argument of this chapter is that there is a need to take (at least) some of the focus off from model-based predictions of future consequences of climate change, to redirect attention towards social and political problems in the present, and to find related ways of embedding the problems within concrete practices and local communities. We argue that, for this to take place, there is a need to reconsider the models of human agency inherent in many analyses of climate change policy, and for integrating broader and more inclusive models of knowledge and agency into the basis for decision making.

Science and the problem of context

Throughout the latter years, many studies in science governance have converged on a focus on the mutual co-production of science and society (Jasanoff, 2004; Latour, 1993). One insight to emerge from such studies is that both the context of knowledge production and the context of knowledge application (Nowotny *et al.*, 2001), including the wider political purposes for which knowledge is constructed, matters more to the ways in which we use science to make sense of the world than is commonly recognised. The implications of this seemingly simple observation should not be underestimated. When considering the uses of science and technology for purposes of broad-scaled political intervention and change, keeping the attention too narrowly focused on the scientific facts of the matter can blind us to the wider context within which we find ourselves.

Similar points have been put forward, although perhaps with less descriptive accuracy, since the inception of the Enlightenment, for instance by Michel de Montaigne and Blaise Pascal. One central point of critique has been, and remains, how the establishment of standardised and impersonal ways of producing knowledge through science also entails the effective *decoupling* of knowledge from the context in which it was created (Toulmin, 2003). It was a decoupling of the means of knowledge from the ends of knowledge that seemed to follow from its universal character and from standardisation, as if offering 'a view from nowhere' (Nagel, 1986).

One outcome of the broad-scaled application of science and technology to nature and society that eventually followed was the radical increase in welfare and prosperity that took place during industrialisation and the coming of the knowledge society. The power of nuclear energy, fossil fuels, antibiotics and computers resides in their enhancement of human agency, i.e. in increased capacities to act in a number of situations. It also resides in the fact that science and technology may be distributed across and used in a number of different contexts, to 'act at a distance' (Latour, 1987). On the negative side, this transformation of knowledge and society also entailed an increasing amount of non-intended side effects to nature and society (Beck, 1992), the unforeseen accumulation of chemicals in ecosystems and the aggregation of CO_2 emissions being prime examples. It also entailed serious regulatory and legal problems: the global, aggregated effects of a number of local actions meant the effective decoupling of responsibility from those effects (von Schomberg, 2007).

Hence, with the decoupling of science and technology from their context of creation, there followed, almost as by logic, the problem of how to re-introduce knowledge into contexts different from the one in which it arose. The problem of the context is at least, therefore, double. On the one hand, we note the dependency of science on the context in which it was created, significantly for its verification and accountability. On the other hand, there is the problem of what to make of science as soon as it is applied for the sake of action, i.e. reintroduced in some practical context. Closely related, there is the seeming impossibility of relating it back to one agent who can be held accountable for negative consequences. Neither of these elements, we argue, is sufficiently attended to in climate change knowledge governance today. We believe such problems of contextualisation to underlie at least some of the broader problems we witness: first, failing efforts to change society towards more sustainable ways of producing and living, second, the related problem of engaging people for the sake of common action towards such a goal.

Problems of science and action in the IPCC policy discourse

It was the creation of the IPCC as such that formally established the global discourse of climate change as a threat to the global environmental system. Prior to this, models of climate would be locally constituted and would, at the most, be of relevance to regional politics. As told by Clark Miller, 'By the early 1980s there was an alternative to viewing climate as merely the aggregation of the weather. Based on computer models of the general circulation of the atmosphere, climate scientists increasingly represented the Earths climate as an integrated, global system' (Miller, 2004, p. 54). Therefore, as far as moral and political implications have been drawn from the findings of the IPCC reports, they remain strongly dependent on the scientific discourse that opened up the perspective of global action in the first place. This is problematic insofar as the seeming certainty about climate change taking place cannot be matched by a corresponding and comprehensive vision for action. In this section we look at how the IPCC reports frame the language of decision making, and how a framework is established for the containment of the problem within the boundaries of climate science, economy and the 'social attitudes towards risk'. We single out three problems related to that framing.

In spite of growing scientific certainty, so far topped by the statement of the 4th Assessment Report that climate change is 'highly likely' caused by human intervention, the necessity of acting under conditions of (scientific) uncertainty has made up an integral part of the policy advice built into the IPCC reports. In the 2001 Synthesis Report it is stated that 'Climate change decision making is essentially a sequential process under general uncertainty' (Watson, 2001, p. 5). This also leads to a recognition of the value-based and political character of many decisions, such as setting the measures for 'dangerous anthropogenic interference' (ibid., p. 35). But rather than letting uncertainty getting the upper hand on science, a conceptual move is made that contains the variables and render the situation manageable: 'Scientific evidence helps to reduce uncertainty and increase knowledge, and can serve as an input for considering precautionary measures. Decisions are based on risk assessment, and lead to risk management choices by decision makers, about actions and policies' (ibid., p. 39). Therefore, even where uncertainty is recognised as immanent to climate change decision making, this uncertainty is of such a kind that the information we gather from models is considered sufficient to represent the general development of the system in question. This way of framing the scientific issues also carries important implications for the ways in which broader policies have come to be defined. The kind of uncertainty that is being recognised is mainly quantifiable, thereby also working to internalise the impacts and effects of variables about which we are ignorant (Hoffmann-Riem, 2002).

What is especially noteworthy for our purposes, however, is the ways in which risk assessment (the paradigmatic case being the graduations of uncertainty given to the scientific findings) and risk management, are translated into the related policy context that is opened up by the reports. Whereas the policy context is recognised as even harder to predict than the climate system itself, also here the number of factors that enter into the equation is limited:

> Decision making has to deal with uncertainties including the risk of non-linear and/or irreversible changes and entails balancing the risks of either insufficient or excessive action, and involves careful consideration of the consequences (both environmental and economic), their likelihood, and society's attitude towards risk.
>
> (Watson, 2001, p. 41)

The main risks faced by policy makers are defined as either doing too much or doing too little. It is hard to disagree with that particular statement. But the language in which such action is framed is problematic, as action is conceived within the conceptual, disciplinary and institutional domains of climate science and economics.

But let us for the moment assume that these disciplines are capable of providing us with more or less accurate predictions of future developments in markets and the climate system. It follows that the results to be gained from such predictions are to be applied by policy makers in the making of scientifically sound politics. Following that, however, further uncertainty factors emerge in the form of 'society's attitude towards risk'. In the part of the 2007/2008 UN Development Report on International Public Opinion, Perception and Understanding of Global Climate Change, it is stated that 'what the public perceives as a risk, why they perceive it that way, and how they will subsequently behave are . . . vital questions for policy makers attempting to address global climate change . . .' (Leiserowitz, 2007/2008, p. 2). This perspective is also important for the sake of evaluating the legitimacy of dominant policies: 'Public support or opposition to proposed climate policies will also be greatly influenced by their risk perceptions' (ibid.). Both the question of predicting social responses as well as the question of political support and legitimacy emerges through the lens of 'risk perception'. All in all, therefore, decision making is seen to consist of the complex interactions of three systems: the climate, the economy and society, which is interpreted as attitudes towards risk. We see three problems with this approach.

First, it translates the risk analysis established for the climate system by the IPCC reports to the understanding of social environments: scientific knowledge about risk remains the main standard for measuring the level and degree of knowledge among populations. For instance, the UN Report builds on questions such as 'Have you ever heard of the environmental problem of global warming?' (ibid., p. 4), and 'How convinced are you that human activities are a significant cause of changes to the Earth's climate and long-term weather patterns?' (ibid., p. 15). Other questions are about the level of concern among population groups across the globe, thus establishing a link between *knowledge about* climate change and *concern* about climate change. All in all, this reinforces the view promoted in the IPCC report of a close connection between perception of risk and prospects for action. This way of linking knowledge with action comes close to what has been labelled the 'deficit model' of public understanding of science, describing public ignorance of science as a main obstacle to successful public policy (Irwin and Wynne, 1996). Relevant means, according to this model, may be information and education campaigns to increase the general level of knowledge among populations.

We do not exclude the possibility that the deficit model, in given contexts, may actually work, and that it may have its legitimate uses, especially within cultures that are scientifically advanced and with high scientific literacy (Sturgis and Allum, 2004). This does not cancel out the many critiques raised against models of agency that take their main clues from natural science and economics (see, for instance, Taylor, 1985, 1989). Instead, it situates knowledge-based action within a problematic field in which scientific knowledge is one among several decisive factors in the decision making of people and communities.

When it comes to prescriptions for action, also the UN Report easily makes the leap from the discourse of science to the discourse of economics: 'The very limited data we do have at least suggests a willingness by many worldwide to pay higher prices for fuel if the money raised was devoted to reducing air pollution and higher prices for electricity produced by renewable energy sources' (Leiserowitz, 2007/2008, p. 36). Thus, in spite of admitting to large gaps in the knowledge base, and in spite of being more informed by social science models, the UN Development Report ends up by affirming the IPCC image of the 'global lay decision maker' as acting mainly according to the prescripts of science and economics.

Hence our argument here is not that people do not understand science, or that climate science should not be communicated as important to the people of the world. The problem is, as mentioned on p. 151, that of re-contextualising, within a myriad of different policies, communities and life-worlds, the global discourse opened up with the IPCC reports. As stated by Stephen Toulmin,

> There is a . . . contrast between our local knowledge of the patterns we find in concrete events, and the universal, abstract understanding embodied in purely theoretical points of view. The substance of everyday experience refers always to a 'where and when': a 'here and now' or a 'there and then'. General theoretical abstractions, by contrast, claim to apply *always* and *everywhere*, – and so . . . hold good *nowhere-in-particular*.
>
> (Toulmin, 2003, pp. 15–16)

For most people, the discourse promoted by the IPCC reports as well as the UN report speaks from this global perspective, from a great distance. That may be fine, as a first approximation, but it needs to be recognised that such discourse-at-a-distance is better suited for purposes of social engineering than for engaging and mobilising people in the places, cultures and governance structures within which they find themselves. Thus, over-hasty applications of the models originating in the IPCC report may easily result in misconceptions about human agency and social action.

Second, but closely related: insofar as the 'risk perception' model remains the main platform for conceiving of the 'social dimension' of climate change, the democratic and communicative challenges connected to large scale restructuring of society are left out. According to one conception of democracy that has become increasingly popular throughout the latter years, '. . . outcomes are legitimate to the extent they receive reflective assent through participation in authentic deliberation by all those subject to the decision in question' (Dryzek, 2001, p. 651). This perspective remains absent in the IPCC reports, as well as the UN Development Report: Citizens, communities and publics are recognised as parts of the problem of achieving sustainability, but they are not included in the creation of solutions.

Clearly, the scope, scale and complexity of climate change issues render the fulfilment of the deliberative ideal a practical impossibility. That, however, does not remove the normative pull of the argument: insofar as 'society' is conceived exclusively through the concept of risk perception there is a lack of appreciation of the enormous democratic challenges involved in large-scale social and environmental change. It is highly problematic insofar as the main premises for decision making are framed by the use of a few perspectives only, and insofar as the main drivers behind social change remain restricted to a few powerful sectors, i.e. international political organs, the energy and industry sectors and the economy. The problem complex should be seen in relation to the previous point, where we noted a lack of appreciation of more locally embedded forms of knowledge to mobilise populations and communities, thereby also conceptualising and including 'thicker' forms of agency and social action.

Somewhat ironically: with the coming of the 4th Assessment Report, in which the scientific certainty is strengthened compared to earlier reports, there seems to be a growing appreciation of the need to include a wider spectrum of issues related to social and political dimensions: 'Responding to climate change involves an iterative risk management process that includes both adaptation and mitigation and takes into account climate change damages, co-benefits, sustainability, equity and attitudes to risk' (IPCC, 2007, 22).[1] Not the least, this change of emphasis reflects the growing acceptance of adaptation strategies as inherent parts of any comprehensive climate change policy, a view that is also increasingly recognised in the environmental science literature (Biesbroek *et al.*, 2009; Dowlatabi, 2007; Goklany, 2007; Klein *et al.*, 2007; Tompkins and Agder, 2005). Adaptation strategies may be seen as more hospitable to some of the issues referred to above: they entail a wider number of disciplinary fields; they include the active collaboration with local communities and knowledge forms, and may therefore also be seen as intrinsically more democratic. Furthermore, the appreciation of identity and belonging as intrinsic to mobilisation of populations and communities that was largely seen as lacking in the global discourse may be spotted on the horizon of increasing numbers of collaborations between governments, academics, local authorities and local communities (Biesbroek *et al.*, 2009; Few *et al.*, 2006).

However, insofar as democracy and participation are concerned, it remains to be seen to what extent such approaches will also be allowed to exert any influence on the overall setting of policy priorities and goals on high political levels: 'The impacts of adaptive measures are most noticeable locally and are generally not designed to contribute to the reduction of GHG in the long run' (Biesbroek *et al.*, 2009, p. 3). Mitigation strategies, however, deploy different disciplinary resources and belong within a different institutional domain:

> In most cases, mitigation strategies are formulated using information from a limited number of scientific disciplines (mainly technology and economics) and are embedded in sectoral policy domains. A typical approach for mitigation strategies is through institutional arrangements, especially for industrialized nations which signed the Kyoto Protocol and formulated specific measures to reach the top-down GHG emission targets of the Protocol, for example, by financially supporting technological development and innovation, or establishing cap-and-trade schemes. Within this context in which

> problems are framed, and governed by, the interest of one single scientific community, quantitative modelling approaches are used to produce the highly specialized and often complex knowledge.
>
> (ibid., p. 2)

We maintain, therefore, that whereas the increased emphasis on adaptation strategies may be seen to introduce more democratic forms of knowledge creation and political participation, the overarching problem related to the creation of more democratic discourses on climate change policies remains an outstanding task. As can also be seen from the above quote from the 4th Assessment Report, both adaptation and mitigation strategies are maintained within the language of risk assessment and management. At the centre of the discourse there are still few openings for conceiving of the great variety of knowledge forms that will be needed, and there is still no way in which knowledge creation is related to issues of democracy.

Third, the above problems should be connected to the intrinsic uncertainties of scenarios based on climate science and economics. We have seen how the IPCC report recognises the irreducible presence of scientific uncertainty, especially relating to science for policy. We also saw how uncertainty is channelled into tools for risk assessment and management as provided by climate science and economics. In science-for-policy these disciplines are used also for modelling future scenarios about which certain and quantifiable knowledge is lacking. In many cases such ignorance is not openly recognised and communicated:

> If modellers are asked for detailed forecasts about what will happen, say, in south-east England in 2060, some feel that it's their job to provide the best available information. Then they report whatever today's biggest computers spit out, even if they know those results are not robust.
>
> (Lenny Smith interviewed in *New Scientist* 6 December 2008)

With increasing political, commercial and public pressures building up around climate science, the danger is increasing that hasty scientific conclusions feed into policy processes demanding fast and safe answers. Policy makers and scientists may jump to premature conclusions leading to locked-in situations where society is committed to solutions that are neither sustainable nor scientifically nor economically viable. Large-scale commitment to biofuels could serve as one example of such a situation. Given that main policy initiatives are restricted to a limited number of powerful agents, scientific uncertainty and ignorance adds a further twist. Where the predicted scenarios fail to materialise, or, worse, where large-scale policy initiatives introduce novel risks to society and the environment, the legitimacy of governments may be at stake. The problem points to a lack of adequate institutions on the interface between science and politics (Beck, 1998; Callon, 2009; Jasanoff, 2004), and this institutional deficit is exacerbated by the growing pressures for immediate and decisive action in the climate change issue. Again, there is a problem of translation, this time from the science context to the policy context, and again we note the danger that the homogeneous and scientific language of the IPCC reports may fail to communicate crucial issues of knowledge and ignorance.

Climate science and political visions of action

Compared to the somewhat precautious statements from the 2001 report quoted above, it is not just the scientific facts that have become more concrete, it is also the call for decisive action. In a speech given to the UN Commission of Sustainable Development, Gro Harlem Brundtland stated that:

> So what is it that is new today? What is new is that doubt has been eliminated. The report of the International Panel on Climate Change is clear. And so is the Stern report. It is irresponsible, reckless and deeply immoral to question the seriousness of the situation. The time for diagnosis is over. Now it is time to act (Brundtland 2007). The problem remains, of course: *how* to act, and *what* to do?

A prominent example on the global scene would be the stance taken by the European Union. Central goals announced by EU is to prevent temperatures from rising more than 2 degrees above pre-industrial levels, that developed countries reduce their CO_2 emissions to 30 per cent below 1990 levels by 2020, and that developing countries be helped to achieve similar reductions through significant transfers of financial resources. Central means for achieving these goals are the establishment of a global carbon market (following the model of the EU Emission Trading Scheme), the establishment of sustainable power generation from fossil fuels (aimed at zero emissions by 2020), and the development of new and clean technologies (European Commission, 2007). In this way, it is projected that the climate challenge be turned into a win/win situation:

> The basic physical inertia of the global climate system means that ignoring scientific warnings will lead to unprecedented, costly and potentially unmanageable consequences. At the same time, there is an opportunity to address climate change, energy security and the current economic recession together.
>
> (European Commission, 2009, p. 14)

A leap is made directly from the gravity of the situation, as defined by climate science, into prescripts for action that will integrate the need for sustainability, energy safety and economic growth. With the coming of the economic downturn, the momentum of two factors can be seen to be given a new twist, possibly marking a new stage in what Morten Hajer described as 'the discourse of ecological modernisation' (Hajer, 1995). A number of different interpretations have been made about the content of that discourse, not the least within the commentary literature to what eventually became the theory *called* ecological modernisation (for an overview, see Fisher and Freudenburg, 2001). However, some common features of the earlier discourse may be outlined: first, it would be positioned against a general storyline of serious threat to the environment, the 'apocalyptic horizon of environmental reform' (Mol and Spaargaren, 1993). Perhaps needless to say, today the main storyline is the global risk discourse opened up by the IPCC. Second, they would prescribe means, or at least an overarching goal, of integrating previously separate spheres of production, action and knowledge, trying to bring these into greater accordance and harmony aiming for sustainable development. In the description of Marten Haajer: 'Global ecological threats such as ozone layer depletion and global warming are met by a regulatory approach that starts from the assumption that economic growth and the resolution of ecological problems can, in principle, be reconciled' (Hajer, 1996, p. 248). Thus, in this earlier version of ecological modernisation, economic growth and sustainability were seen as two partly opposing values, to be balanced by institutional change and political artisanship. Increasingly, they are presented as two sides of the same coin, 'green development' *being* the answer to the economic crisis: '. . . staying below 2°C will require significant financial resources for emission reductions and adaptation, but . . . this will also stimulate innovation, economic growth and lead to long-term sustainable development' (European Commission, 2009, p. 12).

Also in the EU vision a catastrophic storyline is introduced in order to establish the gravity of the situation: it refers to the 'unprecedented, costly and unmanageable consequences' of inaction. With the recent economic downturn another storyline is also introduced, one of economic stagnation, rising unemployment and collapsing markets. All in all, therefore, we seem to be increasingly faced with the challenge, not of balancing competing sectorial interests, but of thoroughly reforming these sectors and setting them on a new course.

The radical twist of the EU vision for action, compared with that of the Brundtland Report, resides in the way in which it inverts the order of argument, the relation of *Realpolitik* and ideal. Whereas the earlier concept of sustainable development could be seen as a regulative ideal, a goal for which to strive in the complex negotiation and balancing of interests, it now seems that it has become constitutive for action. It is based on the projection that future technological developments and restructurings of markets will provide clean energy, sustainability and economic growth. Hence, a lot comes to hinge upon the successful implementation of measures such as CO_2 capture and sequestration, clean energies (among which atomic energy is emerging as increasingly central) and the establishment of a global CO_2 emissions trading market.

It cannot be known that such goals will or can be met; predictions are made using computer-based models, such as the POLES and the GEM-E3 model (Institute for Prospective Technological Studies, 2005). These incorporate most of the shortcomings noted about climate models and add some: only a limited set of variables is allowed, and highly disputable assumptions about the behaviours of markets, people and climate are built into their basic structures. For instance, the successes of the CO_2 emissions trading scheme was based on the general assumption of a steadily growing market. With the coming of the economic recession, however, that assumption is no longer valid, resulting in slumping prices on the global CO_2 market, which may now be brought near to collapse. According to *The Guardian*:

> As recession slashes output, companies pile up permits they don't need and sell them on. The price falls, and anyone who wants to pollute can afford to do so. The result is a system that does nothing at all for climate change but a lot for the bottom lines of mega-polluters such as the steelmaker Corus: industrial assistance in camouflage.
>
> (Glover, 2009)

Although it may be too early to predict the death of the climate emission trading market, this example, placed in the context of the radicalised EU vision for action, illustrates the problems noted in the previous section. First, most of the solutions propagated by the EU take their major premises from the global discourse of climate change established by the IPCC. However, this entailed a transfer from one domain of knowledge, that of climate science, into the world of political action. We argued that practical solutions to climate change necessarily must incorporate different knowledge strategies from the ones that were used to identify the problem in the first place. Notions of risk or social agency predominant in the IPCC policy vision cannot, without further ado, be transferred into prescriptions for action, especially not in the guise of computer models. Second, the main initiatives promoted by the EU are kept within strict domains of policy, sector interest and knowledge production. Especially where policies promoted by such strong power/knowledge constellations fail, difficult issues relating to legitimacy, trust and democratic viability of government arise. Third, these problems relate to the ways in which knowledge, risk and uncertainty are used and distributed. The handling and communication of scientific uncertainty is important to both scientific and political legitimacy. Failure to establish viable and realistic visions for action point to the 'institutional void' (Hajer, 2003) residing in the spaces between modern science and politics (Latour, 1993, 1998). This open space is further accentuated and expanded through the intensifying global discourse on climate change.

Concluding suggestions: towards an ethics of knowledge and action

Throughout our analysis in this chapter, we have presented the climate change issue and the problem of lack of effective action, indeed a lack of agency, as a problem of *Verfremdung* – alienation. The issue is defined in a scientific context that is perceived as distant from and alien to the communities in which people live their lives and the public spheres in which they exert their citizenship. However, there seems to be no solution to the issue without the radical involvement of citizens. In the absence of a technical fix, climate science prescribes emission cuts that amount to civilisation change. However, the impossibility of eradicating uncertainty undermines the force required of climate change to act at such a distance – from its academic context to global politics, consumption patterns and lifestyles – by mere words. Science is not enough.

It is no surprise, then, that governance bodies such as the European Union try to produce a space for action by redefining the issue in terms of the market and hence reframing it as (also) an opportunity for business and economic growth.

As noted above, it is too early to know the fate of the emission trading schemes. We do not expect all our readers to share our pessimism with regard to such schemes that could be classified as belonging to 'the industrial regime' (Latour, 1998). As social scientists, however, we should contribute to the development of alternative framings of the issue, even if only as contingency measures.

From our argument it will be no surprise that we will advocate a search for such reframings in the direction of a democratisation of the climate issue. We do so because of the danger of policy vacuums that may produce a state of exception; but also because we believe it is a necessary element towards effective climate measures. The very difficult question is naturally what would constitute such a democratisation and de-alienation of the climate issue.

Two apparently deficient routes towards democratisation of climate change are instructive. First, right-wing political parties in several countries have managed to 'solve' the problem of alienation in the climate change issue by aligning with climate sceptics. They have displayed the lack of full scientific certainty and lack of full scientific consensus on one hand, and noted the life-world fact that there is no immediate climatic catastrophe. In this way, they have been able to establish dialogue and alliance with large groups of voters. The implication in terms of action is of course that there is no justified need for dramatic change.

Next, also in the absence of the immediate climatic catastrophe, governmental and non-governmental environmentalist campaigns and organisations have been able to achieve a certain acceptance among citizens for some energy- and emission-reducing changes in behaviour and lifestyle. In Europe, for instance, certain products are becoming illegal (such as incandescent light bulbs), certain habits are becoming usual (such as recycling of glass and paper) and some habits are becoming slightly more frequent (such as abstaining from private cars). In sum, however, the impact of these changes is quite small, if not marginal. We dare to speculate that such initiatives to change will remain ineffective as long as their justification lies in a science-based, hypothetical scenario of the future.

Problems to engage and mobilise populations for the sake of sustainability and change may be more closely related than commonly recognised to policies in which access to participate in the search for solutions are withheld. Why should people engage in issues in which they are offered few or no possibilities to influence the outcome, and that are nevertheless framed in terms that make little sense within their daily lives?

We could broaden the perspective taken by the EC arguing that climate change poses the possibility of establishing more sustainable societies and to rethink democracy at the same time. The need for change could, in itself, pose a starting point good as any. The potential for democratic, technological and social innovation through networks stretching across traditional social and disciplinary boundaries could be the target of mapping, intervention and dialogue. Here, the social and cultural sciences could have important roles to play. Given the fact that climate change policy concerns us in a great many aspects of our lives, and cuts across a great number of socio-economic sectors, essential issues are left out by the almost exclusive emphasis on economic models and risk perception analysis. 'Sustainable policy' is also a question of engaging with emerging trends and movements. Thus, we need not only one-size-fits-all analyses targeted at national or global populations, but qualitative analysis of networks and movements through which new trends, technologies and lifestyles may spread. The kind of social science we are talking about would not be geared towards social engineering, but rather at identifying, deliberating and engaging with emerging popular initiatives, businesses, engineers, scientists and local communities (see, for instance, Callon, 2009).

An ethics of knowledge and action accordingly should begin in the commitment to ascertain what is happening in the present rather than what may happen in the future. The point is that quite a lot of current events and actions can be judged in moral terms without recourse to abstract climate or economic models: excessive use of natural resources for luxury purposes; the accelerated spending of limited and non-renewable resources of energy and materials; the extinction of animal and plant species; the spoiling of wilderness; the maintenance of global inequity of a vast scale; chemical pollution of the soil, etc. If one asks why such practices take place and indeed are allowed to take place, the answer appears to rely on some kind of differentiation and alienation, by which the public is told that there is no practical alternative, that the wrong-doing is justified by the necessities of *Realpolitik* or the Market, and that anyway the situation is under control. It seems that the kind of democratisation we are in search of should begin with the cry that the Emperor has no clothes, that the wrong-doing is morally unjustified and that nobody is in control. In this way, there seems to be a lesson to learn from the right-wing political parties that deny climate change. Rather than with scorn and scientific arrogance, they should be met with concrete evidence relevant to the moral debate that currently is lacking. The outcome of such a process, if at all possible, will be unpredictable: our societies may choose the path of non-solidarity and non sustainability, and it may be that the measures encountered will be insufficient. For pessimists such as these authors, however, the mere chance of a path of solidarity and sustainability is already an improvement of the current situation.

Notes

1 Similar arguments were made in the 3rd Assessment Report. Our point is that the argument has been more widely appreciated, as witnessed in increasing attention to the problem within the environmental science literature.

References

Agamben, G. (2005). *State of Exception*. Chicago, London: The University of Chicago Press.

Beck, U. (1992). *Risk Society*. London, Thousand Oaks, New Delhi: Sage.

Beck, U. (1998). *Democracy Without Enemies*. Cambridge: Polity Press.

Biesbroek, G.R., Swart, R.J. and Wim G.M. van der Knaap. W.G.M. (2009). The mitigation–adaptation dichotomy and the role of spatial planning. *Habitat International*, **33**, 230–237.

Brundtland, G.H. (2007). Speech at the UN Commission on Sustainable Development. Last accessed 30 August 2009 at URL: http://www.regjeringen.no/en/dep/ud/selected-topics/un/Brundtland_speech_CSD.html?id=465906

Callon, M. (2009). *Acting in an Uncertain World: An Essay on Technical Democracy*. Cambridge: The MIT Press.

Dowlatabi, H. (2007). On integration of policies for climate and global change. *Mitigation Adaptation Strategy for Global Change*, **12**, 651–663.

Dryzek, J. (2001). Legitimacy and economy in deliberative democracy. *Political Theory*, **29**(5), 651–669.

European Commission (2007). *Limiting Global Climate Change to 2 degrees Celsius. The Way Ahead for 2020 and Beyond* (COM(2007) 2 final), ed. Commission of the European Communities. Brussels: Commision of the European Communities.

European Commission (2009). Towards a comprehensive climate change agreement in Copenhagen (COM(2009) 39 final), ed. Commission of the European Communities. Brussels: Commission of the European Communities.

Few, R., Brown K. and Tompkins, E.L. (2006). Public participation and climate change adaptation. Tyndall Centre Working Paper No. 95, Tyndall Centre for Climate Change Research.

Fisher, D.R. and Freudenburg, W.R. (2001). Ecological modernization and its critics: assessing the past and looking toward the future. *Society and Natural Resources*, **14**, 701–709.

Giampietro, M. (2008). Socio-ecological reconstruction: using a new analytical approach to explore the option space and the dialectical tension implied by sustainability, manuscript.

Glover, J. (2009). Carbon market crashes with a whisper, not a bang. *The Guardian*, Friday, Feb 27. London.

Goklany, I.M. (2007). Integrated strategies to reduce vulnerability and advance adaptation, mitigation, and sustainable development. *Mitigation Adaptation Strategy for Global Change*, **12**, 755–786.

Hajer, M. A. (1995). *The Politics of Environmental Discourse: Ecological Modernization and the Policy Process*. Oxford: Oxford University Press.

Hajer, M. A. (1996). Ecological modernisation as cultural politics. In *Risk, Environment and Modernity, Towards a New Ecology*, ed. S. Lash, B. Szerszynski and B. Wynne. London: Sage.

Hajer, M. A. (2003). Policy without polity? Policy analysis and the institutional void. *Policy Sciences*, **36**, 175–195.

Hoffmann-Riem, H. and Wynne, B. (2002). In risk assessment, one has to admit ignorance. *Nature*, **416**(6877), 123.

Institute for Prospective Technological Studies (2005). Analysis of post-2012 climate policy scenarios with limited participation. In Technical Report European Commission DG Joint Research Centre.

IPCC (2007). 4th Assessment Report. Climate Change 2007: Synthesis Report. Summary for Policymakers. The International Panel on Climate Change.

Irwin, A. and Wynne, B. (1996). *Misunderstanding Science? The Public Reconstruction of Science and Technology*. Cambridge: Cambridge University Press.

Jasanoff, S. (2004). *States of Knowledge: The Co-Production of Science and Social Order*. London: Routledge.

Klein, R.J.T., Huq, S., Denton, F., Downing,T.E., Richels, R.G. and Robinson, J.B. (2007). Inter-relationships between adaptation and mitigation. In *Climate Change 2007: Impacts, Adaptation and Vulnerability*. Contribution of Working Group II to the Fourth Assessment Report of the Intergovernmental Panel on Climate Change, ed. M.L. Parry, O.F. Canziani, J.P. Palutikof, P.J. van der Linden and C.E. Hanson. Cambridge: Cambridge University Press.

Latour, B. (1987). *Science in Action*. Cambridge, MA: Harvard University Press.

Latour, B. (1993). *We Have Never Been Modern*. New York: Harvester Wheatsheaf.

Latour, B. (1998). To modernise or ecologise? That is the question. In *Remaking Reality: Nature at the Millennium*, ed. B. Braun and N. Castree. London: Routledge.

Leiserowitz, A. (2007/2008). International public opinion, perception, and understanding of global climate change. United Nations Human Development Report Office.

Miller, C. (2004). Climate science and the making of a global political order. In *States of Knowledge: The Co-production of Science and Social Order*, ed. S. Jasanoff. Abingdon, UK: Routledge.

Mol, A.P.J. and Spaargaren, G. (1993). Environment, modernity and the risk society: the apocalyptic horizon of environmental reform. *International Sociology*, 8, 431–459.

Nagel, T. (1986). *The View from Nowhere*. Oxford: Oxford University Press.

Nowotny, H., Scott, P. and Gibbons, M. (2001). *Re-Thinking Science. Knowledge and the Public in an Age of Uncertainty*. Cambridge: Polity Press.

Sturgis, P. and Allum, N. (2004). Science in society: re-evaluating the deficit model of public attitudes. Surrey Scholarship Online.

Taylor, C. (1985). What is human agency? In *Philosophical Papers*. Cambridge, New York, Melbourne: Cambridge University Press.

Taylor, C. (1989). *Sources of the Self. The Making of the Modern Identity*. Cambridge, MA: Harvard University Press.

Tompkins, E. L. and Agder, N. W. (2005). Defining response capacity to enhance climate change policy. *Environmental Science and Policy*, 8, 562–571.

Toulmin, S. (2003). *Return to Reason*. Cambridge, MA, London: Harvard University Press.

von Schomberg, R. (2007). From the ethics of technology towards an ethics of knowledge policy and knowledge assessment. Brussels: European Commission Services Working Paper.

Watson, R.T. (2001). *Climate Change 2001: Synthesis Report*, ed. Robert T. Watson: The International Panel on Climate Change.

10 Technological idealism

The case of the thorium fuel cycle

Karl Georg Høyer

The historical context

The most critical issues raised by nuclear power use have a global reach and are just as crucial today as they were some 30 years ago. In the 1970s and 1980s, even Norway, a country with vast hydro-power resources and production, was subject to serious nuclear power development planning. According to these plans, Norway by now should have had some 12–15 nuclear reactors localised to 4–5 nuclear power plants. Due to strong public opposition this was, at least initially, rejected by the Norwegian Parliament as early as 1975. Similar plans were also rejected in Denmark and, ever since, the two Nordic countries have kept their roles as nuclear power free zones. In Sweden, the Parliament decided gradually to dismantle and phase out all their existing nuclear reactors, a decision very much highlighted in broader international discussions. However, in recent years it has proven difficult for Sweden to keep to the decision, when confronted with the issue of climate change (Høyer, 1977).

Two major nuclear reactor accidents should heavily influence both discussions and decisions. The first one was in March 1979. A loss-of-coolant accident (LOCA) took place in one of the two reactors at the Three Mile Island nuclear power plant near Harrisburg in Pennsylvania, USA. Before control was regained, the reactor was only a few hours from a fuel meltdown accident. 140,000 people had to leave their homes for a shorter or longer time. In its effects, uncontrolled emissions of radioactivity to the ambient environment, it was not a serious accident, but it demonstrated all the potentials of the utmost severity. And not least it demonstrated the necessity to throw all former quantified risk estimates into the garbage can. In the USA the accident lead to a moratorium in commissioning new nuclear reactors. It took 6 years before the Three Mile Island reactor could start up again, strong evidence for the vulnerability of nuclear power as an energy source if larger or minor accidents happen (Brøgger and Høyer, 1986).

Based on the almost unanimous recommendations from a public commission, some industrial activists in the mid 1980s made efforts to restore nuclear power planning in Norway. Their timing was not very good, at least not for themselves. On 26 April 1986, reactor 4 in the Ukrainian Chernobyl power plant became subject to the most severe nuclear power accident in history. As widely recognised, extensive land areas and populations both in Ukraine, Belarus and Russia were particularly seriously hit by radioactive fallout. But even as far away as the more remote parts of Norway, the fallout was large enough to make immediate counter-measures necessary in order to protect the population from long-term health effects. Now more than 20 years later some of these counter-measures are still effective, in particular those that were enforced to counteract radioactive cesium concentration in reindeer and sheep meat generated through mountain grazing. The total fallout of Cesium 137 and 134 over all of Norway was not a large volume in common terms. In theory, it could be kept in a tea cup. On the other hand, the amount of radioactivity was very large indeed, and at fairly elevated levels is estimated to be present in Norwegian ecosystems through most of this decennium.

Together with other European countries, Norway was taken by surprise in many ways. First of all, it was the large geographical outreach of quite a substantial fallout. Almost all European countries became victims, many subject to heavily concentrated fallout at very large distances from the Chernobyl source. These patterns and distances of radionuclide spreading were very different from the existing models used in risk estimation and contingency planning. And there were all the biological concentration chains of the radionuclides, many never envisaged before, at least not with regard to their proven importance. Former models and estimates of biological halftimes of nuclides were rejected by hard evidence from the 'open systems' of reality.

Then there was the accident itself. Most European and American experts – this author included – shared the view that the Russian graphite moderated RBMK reactors were inherently less accident prone than the Western light water moderated reactors (LWR), whether of the pressurised or boiling type. It was generally accepted, in principle, that the LWR reactors could be subject to a total, uncontrolled meltdown accident, contrary to the RBMK technology. This was the so-called 'China syndrome', hot fuel melting its way down in the ground visually towards China from the USA. We were all surprised by the type and extent of the Chernobyl reactor accident, but not by the release of radioactivity when the accident took place after all. Of course, the lack of the external safety barrier in most Russian reactors at that time, so crucial in Western reactors, was heavily criticised (Brøgger and Høyer, 1986).

This chapter considers the major problems with nuclear power as they have been outlined since the early 1970s. Apart from the accident hazards, there are the safety and deeply ethical problems raised by the continuous generation of long-life radioactive waste, and similar types of problems caused by long-term decommissioning of various types of nuclear fuel cycle plants, reactors, reprocessing and enrichment units. There are the transport safety issues of linking all the fuel cycle plants and activities together. And not least, there is the inherent and potentially serious connections between nuclear power and nuclear bombs, where the very history of nuclear power was founded more than 60 years ago. Nuclear reactors are still continuously generating plutonium-239, the isotope applied in the Nagasaki nuclear bomb. And uranium enrichment facilities are creating opportunities for the generation of sufficiently enriched uranium-235, the isotope applied in the Hiroshima nuclear bomb.

In Norway, as well as in other European countries, the strong opposition against nuclear power caused energy issues to be focused in new ways. The large potentials of new forms of energy production from renewable and environmentally benign sources, sun, wind, biomass and low temperature heat from the ground, have become crucial parts of the new way of thinking on energy – the *soft energy paths* (Lovins, 1977). And in outlining these paths, extensive energy saving and gains in energy efficiency have come to play important roles. In the aftermath of the nuclear power opposition and the many no's to further nuclear power, it was mostly a matter of change of focus in discussion, but gradually it has grown to be an integral part of new politics. And, as the future is described today, with the overriding response to issues of climate change, the soft energy paths are taken more seriously than before in all European energy developments and policies.

The climate change issue, however, has caused nuclear power to re-enter the political agenda in Norway as elsewhere. Internationally, the new term applied is *sustainable nuclear energy* (Wise, 2005; Samseth, 2007). A new nuclear technology, even termed the *Norwegian solution*, was launched with a lot of enthusiasm in 2005 as an ultimate form of sustainable nuclear energy. Popularly it was known as *thorium reactor technology*, and thus implied use of thorium instead of uranium as a fuel resource. But it was not any type of thorium reactor. The technology proposed had, as its most crucial component, a *particle accelerator*, external to the reactor itself. It was an external source of high energy protons, an energy amplifier, to generate the necessary free neutrons within the reactor configuration. These neutrons would transform the fertile thorium-232 into highly fissile uranium-233 in the reactor core. The scientific term applied to the whole system was *ADS – Accelerator Driven Systems* (Rubbia, 1996; Rubbia *et al.*, 1997; WNA, 2003).

These envisaged plans will heavily influence the Norwegian discourse on energy and nuclear power, at least for a few years. This chapter is about this particular turn in the Norwegian discourse. The critique carried out is based on analyses of texts. However, the term *text* is quite wide. It includes lectures, power point presentations, and articles and interviews in newspapers and on the internet. I limit these to texts produced by crucial scientists, mostly Norwegian university professors, among other claims they have made when interviewed in various media. These claims came to constitute the real basics of the formative process of the new thorium power discourse.

Excursus I: Critical realism in the theory of science

Critical realism (CR) is held as a position within theory of science. It can be distinguished from other such main positions, as positivism, empirical realism, hermeneutics, constructivism and others. In the history of science theory it was initially developed in opposition to positivism in the 1960s and 1970s, however, including basic realism as does positivism. The British philosopher of science

Roy Bhaskar has made the most significant contribution to this position, in which he still is active (Bhaskar, 1979, 1991, 2008). Several other major contributions are published in a book of essential readings (Archer *et al.*, 1998). A dictionary outlining key concepts and definitions was first published in 2007 (Hartwig, 2007). Highly readable Nordic presentations are published in Swedish (Danermark *et al.*, 1997) and Danish (Buch-Hansen and Nielsen, 2005).

CR highlights meta-theoretical and, particularly, ontological issues in the theory of science, and elaborates on the distinctions between ontological, epistemological and methodological questions. *Ontology* encompasses theories on the most decisive aspects of questions of being and of reality. A crucial issue is to what extent reality exists independent of our ideas or knowledge about it. *Idealism* is a position claiming there is no such independence, while an independent existing reality is at the basic of *realism*. *Epistemology* covers theories of knowledge and knowledge production. Crucial issues are about the possibilities of gaining knowledge about a reality and the conditions for knowledge production to take place. *Methodology* is about how actual science is carried out. This may cover abstract issues about what science is, or is not, but also more concrete issues on the very practice of scientific efforts. The next level in this structure is then about the actual *methods* applied. CR is a position emphasising *method-pluralism*, thus opposing method-imperialism, however without implying method-relativism.

CR ontology makes a clear distinction between reality – what it is – and our knowledge about this reality. Science has two distinctly different dimensions. The *transitive dimension* covers our knowledge about reality. Epistemology belongs to this dimension. In actual science, the transitive objects are the paradigms, theories, concepts, models, etc. that are present at a given moment. By several critical realists, they are termed 'the raw materials' of science, creating the indirect links between science and reality. But science produces knowledge about something. This is the *intransitive dimension*, the objects that science aims at acquiring knowledge about. Ontology belongs to this dimension. As emphasised above, CR is a position within realism by claiming that the intransitive dimension exists independently of our knowledge about it (Buch-Hansen and Nielsen, 2005).

A crucial claim in CR ontology is that reality consists of *three separate domains*, a claim that really makes a difference from other forms of realism. The three domains are: the *empirical*, the *actual*, and the *real*. The *empirical* domain covers the observations and experiences we make, be they direct or indirect. It is to be separated from *the actual*, the domain covering all the manifest phenomena existing and events happening in real life, whether we experience them or not; what manifests itself happening in the world is not totally described through what we experience. The third domain is *the real*, based on the claim that all the manifest phenomena and events do not turn up accidentally and all by themselves. The real covers the underlying structures and mechanisms that under certain conditions support or cause actual phenomena and events to take place. They are out of the reach of direct observations and experiences, but are still present in the real world. Table 10.1 visualises the essentials of the three levels of

Table 10.1 The three ontological domains in critical realism

	Empirical domain	*Actual domain*	*Real domain*
Experiences, observations	X	X	X
Phenomena, events		X	X
Structures, mechanisms			X

Adapted from Bhaskar (2008).

ontology in CR. It can also be visualised as *an iceberg*; where the empirical is the very tip, being above sea level, while the real is the large and heavy volume beneath.

As for the difference between the transitive and intransitive dimension, this basic ontology is considered to be common both for humanities, social and natural sciences. However, this claim does not entail the *universalism* ambition in positivism. A common basic ontology is certainly not the same as claiming a common epistemology or a common methodology.

It should be evident that CR is a truly *anti-reductionism* position. The intransitive dimension cannot be reduced to the transitive, or the real cannot be reduced to our knowledge about it. Further, the real cannot be reduced to the actual, even though they are interconnected. There is an *ontological divide* between the two domains. Similarly, there is an ontological divide between the actual and the empirical domain, and the actual can not be reduced to the empirical, even though there clearly are interconnections. In this context CR balances in between two extreme positions of fallacy; on the one hand, the epistemic and, on the other, the ontic fallacy. With the *epistemic fallacy*, ontological questions are reduced to matters of epistemology; questions about reality are reduced solely to questions about our knowledge or discourses about this reality, which actually implies erasing the intransitive dimension of reality and that all three domains, empirical, actual and real, are reduced to only one. The position within realism termed *empirical realism* carries this epistemic fallacy. *Positivism* is not dominated by such empirical realism. With the *ontic fallacy*, the transitive dimension is, so to speak, dissolved. It is the position claiming that reality speaks for itself and can be interpreted as an open book; knowledge is considered to follow directly from reality and questions of knowledge are thus reduced to questions of reality (Buch-Hansen and Nielsen, 2005).

Positivism also entails *empirical causality*. CR shares this interest in terms of causality and causal powers. But this is where the similarity ends. CR has a completely different understanding of how causality comes into play and how it can be addressed in actual science. This leads us to two more concepts in the CR ontology. They are *open* and *closed* systems. In closed systems, empirical regularities may exist. In open systems, however, empirical regularities are generally non-existent. The real always consist of open systems. In order to achieve empirical regularities in science, open systems have to be artificially closed, by man. Most of the science since the advent of modernity has been developed within a closed system paradigm, not only in natural sciences but also in social sciences. Positivism is a position which can only be defended within this paradigm. The CR claim is that all real systems are open systems. One cannot expect empirical regularities when searching for causal powers. In open systems, the causal powers will be present as mechanisms and structures, and cause tendencies to turn up.

A thorium Klondike

> Norway has the fourth largest thorium reserves in the world, totalling 180,000 tonnes. According to the current oil price they represent a value of about 250,000 billion US dollars, or a thousand times the value of the Norwegian oil fund.
>
> (Lillestøl, 2006a)

This claim, made by the Norwegian professor in particle physics Egil Lillestøl, was presented in several texts, in newspapers, in lectures, and on the internet. It will come to be an important basis for the special Norwegian thorium Klondike, notably expressed by the Norwegian company, Thorium Norway Ltd, after having secured themselves the mining rights to a part of the major Norwegian thorium resources located to the Fen complex in Telemark county:

> With the large reserves that are found in our part of the Fen complex area, we could probably deliver a fuel guarantee for more than 1000 reactors during hundreds of years.
>
> (Standeren Pedersen, 2007)

The foundations for these claims are, however, rather shaky, to say the least. One thing is the way the economic value calculations are made. The total amount of electricity generated from reactors utilising all the 180,000 tonnes of thorium was multiplied by the current average market price for electricity. But this has very little relevance to the market price of thorium and thorium fuel. Fuel costs only take up a minor part of the total costs of electricity production from nuclear reactors. In particular, this is the case for the highly capital intensive thorium reactor technology, the particle acceleration technology, proposed in the Norwegian context. Admittedly, it was accepted that the value calculations could be somewhat exaggerated. But as those responsible said: 'It was a matter of attracting potential investors' (Lillestøl, 2006b). And he certainly seemed to succeed. Two new Norwegian companies were established: Thorium Norway Ltd and Thor Power Ltd. After all, the heavy metal thorium was first found in Norway, and was named after the Nordic ancient god 'Thor', the god of thunder with his hammer.

More important is the question to which extent it is reasonable to apply the term *reserve* to these thorium resources. A main reference is the estimate (see Table 10.2) made by the US Geological Survey (2007).

Table 10.2 World thorium reserves

Country	*Th reserves, proven (tonnes)*	*Additional Th reserves, estimates (tonnes)*
Australia	300,000	340,000
India	290,000	300,000
Norway	170,000	180,000
USA	160,000	300,000
Canada	100,000	100,000
South Africa	35,000	39,000
Brazil	16,000	18,000
Malaysia	4500	4500
Other countries	95,000	100,000
Sum total	1,200,000	1,400,000

From US Geological Survey (2007); see also OECD/NEA (2001).

Whether Norway has large thorium resources is not up for question. The crucial term, however, is *reserves*, a term implying resources that are proven or estimated to be technologically and economically available. Even though the international nuclear energy authority (IAEA) has published similar estimates, the scientific substantiation remains poor or even non-existent, according to the Norwegian Geological Survey (GEO, 2007).

The Norwegian thorium resources, however, are not readily available. In a global context, heavy mineral sands, mostly beach sands, are an important source of thorium. Such thorium resources having significant size and grade are not known in Norway, neither as sands nor as metamorphosed sands. In Norway, local ore concentrations are relatively low, and the fine-grained structure causes large technological and economical barriers for the necessary enrichment processes. Per unit weight, the concentrations in the Fen complex, at most, may be about 0.4 per cent, while it may be above 10 per cent in the Indian and Australian monazite-bound sand resources. The fine-grained structure makes established flotation enrichment technologies difficult to apply.

As long ago as the early 1970s, a Norwegian report made this conclusion about the availability of the thorium resources: 'The minerals are so fine-grained that they can not be enriched with satisfactory recovery using traditional techniques from the 1960s and 1970s' (Megon, 1973). More advanced methods for mineral separation must be utilised and adapted to the specific deposits. Separation of very fine-grained minerals as these, however, is a very great technological and economical challenge (Dahlgren, 2008; Kara *et al.*, 2008).

Vast and much more readily available thorium resources are found both in India and Australia. Besides the concentration and enrichment assets, they only require open pit mining. In the Norwegian Fen complex, underground mining is required, adding to the economical barriers. Thorium recovery and fuel production is, of course, a global industry. In this context, the Norwegian resources are not particularly interesting. It is not substantiated that they have any economical value at all. Scientifically it is just as fair to claim the real economical value to be negative.

Liberation from the nuclear bomb legacy

> It is almost technically impossible to produce nuclear weapons with this technology. There is so little plutonium coming out that one in that case would have to collect such waste for 20 years.
>
> (Lillestøl, 2006c)

The resource base for the thorium reactors is Th-232 and consequently U-233. As a point of departure this does not lead to production of Pu-239 in any amounts, simply because the number of neutron absorptions required is too large. Pu-239 is a major nuclear waste product from the common uranium reactors, and is also known as an important material for nuclear bombs. Enriched U-235 is the other major nuclear bomb material, and is a nuclear fuel for the uranium reactors. Sufficient enrichment for bomb purposes requires special enrichment technologies, while the production of Pu-239 only requires reactors based on fuel with a much lower enrichment grade.

Th-232 is not in itself a *fissile* resource. Similar to U-238, however, it has *fertile* capabilities. After neutron absorption in the reactor, it is transformed to the fissile U-233, which is the burner fuel in a thorium reactor. Actually, the term thorium reactor is somewhat misleading. As other reactors, it is really a uranium reactor, but with the isotope U-233 as the fissile fuel. With thorium as the raw material, these reactors continually produce large quantities of the new fissile material U-233.

U-233 can also be applied as a bomb material. As in the case of Pu-239, it does not require any enrichment facilities, only reactors and fuel reprocessing. In its pure form, U-233 has actually very good qualities as a bomb material, and is on the official IAEA-list of such materials together with U-235 and Pu-239. But U-233 is not readily available in a pure form. During the reactor process, it becomes 'polluted' with the isotope U-232 and its daughter products Bi-212 and Tl-208 (bismuth and tallium), both hard gamma emitters. This is exactly the form of 'pollution' that makes reprocessing of used thorium fuel a really hard endeavour (Standring *et al.*, 2008).

Accepting these physical limitations, it's still possible to use reprocessed U-233 for bomb purposes. Thus large quantities of a new nuclear bomb material are produced and possibly put into international circulation. The thorium reactors themselves may also be misused to produce Pu-239. Just as the particle accelerator technology can be applied to transmute radioactive waste with particularly long halftimes, it can also be used to produce Pu-239 with natural uranium instead of thorium as the fuel load. It is fair to conclude that development of large-scale ADS thorium reactors will contribute to new risks for proliferation of nuclear bombs internationally. This is a conclusion supported by both the two public Norwegian commission assessments of the thorium nuclear alternatives (Kara *et al.*, 2008; Standring *et al.*, 2008).

Excursus II: Complexity, pooled interdependence, vulnerability and thorium gated communities

In a real world of *open systems*, complexity and vulnerability are interconnected. And increase in vulnerability makes systems more accident prone; however, not necessarily with subversive external effects.

This is fundamental in the theory of 'normal accidents' developed by the American organizational sociologist Charles Perrow (1984, 1999, 2007). He bases his theory on empirical analyses of accidents and risks in a number of various organisations and technical systems, nuclear power among them. The theory is generic, applying both to technical, societal, organisational and institutional systems and vulnerabilities.

As a point of departure, he launches two basic but interrelated dimensions, each consisting of two contrasting concepts. They are *linear–complex interactions*, on the one hand, and *loose–tight couplings* on the other. *Linear interactions* Perrow connects to expected and well-known sequences of production and maintenance, and the types of sequences that are readily visible, even though they may be unplanned. Complex interactions on the other hand comprise unknown sequences, or unplanned and unexpected sequences, which either are not visible or immediately understandable. Loose and tight couplings are terms well known from engineering sciences. *Couplings are tight* when the buffers or slacks between system components are small; an incident in one immediately and directly affects the other components. *Loosed coupled systems*, on the other hand, can readily incorporate errors, shocks and pressures of change without a resulting destabilisation.

Figure 10.1 illustrates the relations between these two dimensions and system vulnerability, on the one hand, and resilience on the other.

One of the basic Perrow claims is that, in a real world of open systems, failures will always turn up as nothing can be perfect, whether we are talking about construction, equipment, operational procedures, operational personnel, materials, supply, safety equipment and control, and not the least the ambient environment and society. If the interactions are complex, failures will often be unexpected and unrecognisable. If the system also is tightly coupled, failures will not be limited to singular components or units, but strike out the whole system. Complex interactions contribute to personnel confusion, and tight couplings cause the failures to propagate faster than remedial counteractions can be taken.

Interactively complex systems are dominated by *pooled interdependence*, a term applied in organizational theory. This is the case when all components – included personnel – must co-ordinate their inputs if the system is to function at all. If one component is malfunctioning or closed down, a lot of other connections are at least temporarily disconnected because the components are interconnected in diverse ways. In such complex systems parts, units and part-systems serve multiple functions. Then they become much more subject to *common-mode failures*, another well-known term from engineering sciences. This occurs when there is a failure in one component serving many functions at the same time; many functions are struck out by one singular component failure.

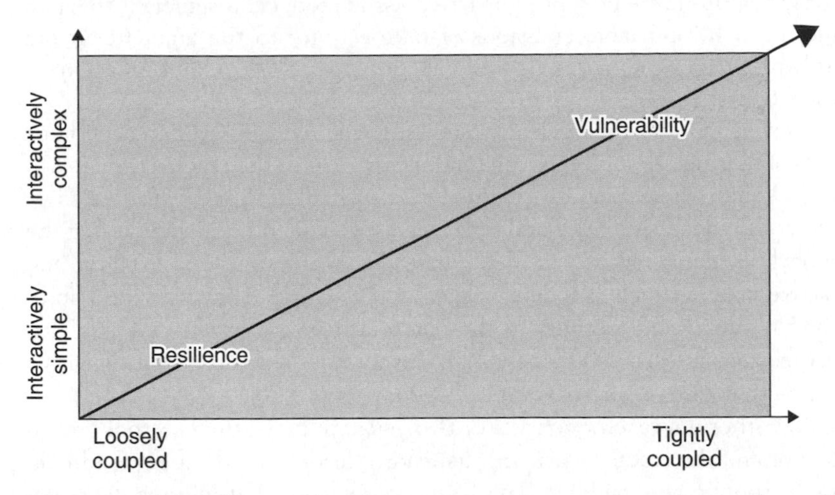

Figure 10.1 Vulnerability and resilience connected to system couplings and complexity (adapted from Perrow, 1984).

Without risks for major radioactive accidents

> It's impossible to have the kind of accident that happened in Chernobyl, simply because the reactor will not melt down like it did then.
>
> (Lillestøl, 2006d)

> A melt-down due to operator errors is not possible.
>
> (Omtvedt, 2009)

The Chernobyl reactor accident was *not a melt-down accident* of the kind discussed internationally since the early 1970s. Some of the fuel melted, but the accident was caused by explosions and fires, not by uncontrolled criticality. The term 'melt-down accident' thus refers to the uncontrolled, chain reaction melt-down of all fuel caused by a total loss of coolant effect. It is a type of accident generally considered possible in all Western type light water reactors (LWR), whether they are boiling water (BWR) or pressurised water (PWR) reactors. Metaphorically, the term 'China syndrome', has been applied, which is also the title of an American film dramatically describing how such an accident could take place. In the film the whole fuel load melts its way down into the ground, visually towards 'China' right at the other end of the world (Brøgger and Høyer, 1986).

This type of accident is not possible in a thorium reactor. There is an effective external control factor available which is not present in conventional reactors. It is the external particle accelerator. In the case of a loss of coolant effect accident, the accelerator can be turned off quite easily, and this will stop a possible fuel melt-down from becoming an uncontrollable chain-reaction. But, a fuel melting accident can still also take place also in a thorium reactor. With a loss of coolant, the decay heat from the present radioactive fissile products is sufficient to cause the fuel to melt, and large emissions of radioactivity to the ambient air are certainly possible if reactor vessels break. In principle, this could be an accident like the Chernobyl one.

The next section on the nuclear waste issue will give an outline of the large amounts of radioactive substances present at any time in the reactor core and in the target liquid, where they are generated both by spallation and by fission. They include both radioactive gases and volatiles as we got to know them after the Chernobyl accident; radioactive noble gases such as xenon and krypton and radioactive volatiles such as iodine, cesium and bromine. They will build up to very high levels of radioactivity in the reactor cover gas system (CGS). This activity may be up to 100,000 times as large as in other reactors operating with a lead–bismuth cooling liquid (Kara *et al.*, 2008, p. 66).

An ADS thorium reactor consists of two separate parts: the external proton accelerator and the reactor itself with its lead or lead–bismuth target fluid held within the containment building. This is an *accident prone* configuration. It creates a vulnerable intersection between the two parts. If there is a rupture in the proton beam tube, or in the interface window between the beam tube and the target fluid, this may break the barrier between the internal reactor system and the surroundings. A misaligned proton beam of possibly 10 MW may burn through the beam tube and also cause vital component melting to take place. The whole configuration is dependent on a highly reliable continuous operation of the accelerator and its beam current. However, *beam glitches*, short interruptions in the beam current, are possible. This will induce very high stress in the intersection window and fatigue of crucial reactor elements (Kara *et al.*, 2008, pp. 55–66).

In themselves both two parts, the accelerator and the reactor, are complex systems together with the beam tube connecting them. Each must be working for the complete system to produce power. There is a higher degree of complexity than in conventional nuclear reactors and power plants. Both basic parts, really all three, need to be turned off for maintenance or when malfunctions turn up. These are reasons to expect lower power production availability than for conventional reactors, even though less frequent shutdown for fuel change may be possible due to longer fuel burn up and smaller criticality fluctuations.

One way or the other, however, there are no operating experiences to substantiate any claims about functionality or availability. In particular, this is the case for the reactor system, where even some basic metallurgical issues remain unsolved at this scale, at least with molten lead as the cooling and target liquid. However, it also applies to the accelerator system. The basic technology here is well demonstrated, but this is only on quite a small scale. The largest accelerator with valuable operating experience today is 1 MW. With the envisaged reactor core of 1500 MW heat production, an accelerator of at least 10 MW is required, an electricity effect presupposing a criticality k-factor of 0.98. (When the k-factor is larger than 1, every fission leads to more than one new fission and a chain reaction takes place, which is the condition for criticality.)

With a larger margin for the *subcriticality*, for instance, a k-factor of 0.95, the required accelerator effect is about 25 MW (Kara *et al.*, 2008, p. 57). When designing a real-life ADS thorium reactor, there will always be a matter of balancing between the degree of subcriticality and safety, on the one hand, and size and, particularly, costs of the accelerator, on the other.

Besides the reactor and its accelerator, the complete ADS power system comprises two other critical components, reprocessing and fuel fabrication. Both are proposed to be integral parts of each power plant in a form of a *thorium gated community*. Such a structure is completely different from others in the whole history of nuclear power development, where reprocessing and fuel fabrication have been large separate industrial plants serving a lot of reactors, for instance, the Sellafield reprocessing plants serving a lot of reactors, not only in Britain but also in the rest of Europe and other continents. *In situ* reprocessing and fuel fabrication will contribute substantially to the total complexity of the thorium power plant, and to its power production costs as the economy of scale is made impossible. Two more complex technological systems are added, and both systems must function efficiently and online with the others. The reprocessing facility has particularly critical functions and aspects. A basically new reprocessing technology is presupposed, termed *pyroelectric reprocessing*. There are no experiences from an industrial scale. Extensive regaining of plutonium and the other minor actinides with an envisaged efficiency of 99.9 per cent adds to the plant complexity and costs, only further to be increased in the case of additional regaining of long-life fission products. A further elaboration is given in the next section on nuclear waste.

A real solution to the nuclear waste legacy

> The amount of waste is much less than when production is based on uranium. Five years production will give about three tonnes of waste, a quantity with a volume of half a cubic meter, but only about 180 kg of this require special treatment and depositing for 400–500 years. A minor amount of plutonium and other transuraniums in the waste are separated and returned to the reactor, which makes the waste much less radioactive. It is almost technically impossible to produce nuclear weapons with this technology. There is so little plutonium coming out that one in that case would have to collect such waste for 20 years. After 400 years thorium-waste is less radioactive than coal ash from the same amount of energy . . .
>
> (Lillestøl, 2006e)

> Waste from a thorium reactor only needs to be stored for 700 years.
>
> (Omtvedt, 2009)

In the fuel load, thorium reactors will generate substantially lower volumes of nuclear waste than conventional uranium reactors. The fuel burn-up is much more complete, thus also giving a much higher energy utilisation of fuel than in the case of uranium. This factor could be as high as 200 per unit weight of fuel. Also achieved is a substantially lower quantity in the waste of plutonium and other actinides with very long radioactive half times. The actinide Pu-239 is known for its radioactive half-life of about 24,000 years, often considered to require a guaranteed isolation for some 500,000 years, 20 times its half-life. In addition, several other actinides are highly problematic with the uranium fuel cycle. Such important assets of thorium reactors are, however, are the case only when a set of rather ideal conditions are fulfilled. These will be critically scrutinised, one by one.

As emphasised, thorium is only a fertile, and not a fissile, resource. To get started, any thorium reactor needs some *topping fuel* of fissile materials. When the whole thorium fuel cycle is in functioning, then the topping fuel will be U-233. But before reaching that stage, the topping fuel must come from the present uranium fuel cycle, most probably U-235. Ensuing neutron absorption in the fuel will result in a waste containing both Pu-239 and the other actinides generated in conventional uranium reactors. Quantities will be relatively smaller, but from many reactors still sufficiently present to add to the already existing volumes of long-life nuclear waste. With the envisaged plans of some 20,000 thorium reactors worldwide, this will increase, not decrease, the total world volumes of long-life plutonium and actinide waste (Martiniussen, 2007; Standring *et al.*, 2008).

In principle, the topping fuel for new thorium reactors can be U-233 generated in the first one's reactors. This is, however, critically dependent on the so-called *doubling time*, the time needed for any breeder reactor to breed twice as much new fissile fuel as it consumes. When this is achieved, the waste fuel has a load of U-233 sufficient to refurnish itself and to start another reactor. A limitation is that the doubling time can be long. It depends on several factors not readily known, as the actual fuel and reactor configurations are not known. It could, however, be a matter of *decades*, though shorter than the hundreds of years needed for other types of thorium breeder reactors. A reasonable conclusion is that, in the foreseeable future, an extensive worldwide thorium fuel cycle only perpetuates the dependence on uranium fuel and the inevitable generation of long-life actinides and nuclear waste.

The plans are critically dependent on highly effective reprocessing. In addition to U-233, all Pu-239 and the other more minor actinides, americium, curium, neptunium, must be totally separated from irradiated thorium fuel, as proposed with losses less than 0.1 per cent. But this reprocessing technology is not available. It is technologically more challenging to reprocess such thorium waste fuel than the current uranium fuel. According to the plans it is proposed to set up individual reprocessing plants connected to each thorium reactor, or to each group of reactors in the case of larger power plants. This will, however, most probably not be accepted by international authorities, simply because it gives each country with such plants the availability to produce their own nuclear bomb material, U-233. Actually, as a means to proliferate more safe nuclear energy technology, the international atomic energy authority, the IAEA, has proposed the very opposite, to establish a global system of central reprocessing plants owned and controlled by international society. The well-proven PUREX technology and its reprocessing plants for irradiated uranium fuel represent the most sad cases in the whole 60 years' history of nuclear energy. A lot of accidents and radioactive leakages have occurred. For Norway, the UK reprocessing plant Sellafield does not give one much confidence, to say the least. Together with the Irish, among others, the Norwegian authorities have continually asked British governments to close the whole plant down. Plans to build a large number of extremely efficient new reprocessing plants based on the envisaged and largely unproven pyroelectric reprocessing technology seem to be unduly idealistic.

After reprocessing and separation, the crucial idea put forward is to recycle Pu-239 and the other actinides and integrate them in a new fuel, together with U-233. For the concept to be complete, this also requires each thorium power plant to have its own fuel fabrication facility. There are reasons to question to what extent this can be realised at all. But let us accept it for the time being. Recycled in the fuel, Pu-239 and the other actinides present will be fissioned and transmuted to stable substances or radioactive substances with shorter half-lives. In this process they will also partly contribute to energy production in the thorium reactor. This certainly is technologically possible. It is a system that also may also be applied to destroy current global bomb material deposits of Pu-239. The idea, however, is far from new. It has been proposed on several occasions for about 40 years, as it is possible to carry out such recycling and actinide transmutation also in conventional uranium reactors. The idea of applying accelerator technology to such purposes has been proposed several times before, but has never been realised. A major reason has been the economy of it – an important part of the real world.

The rest of the radioactive waste from the thorium fuel cycle contains all the major forms of radioactive fission products known from the conventional uranium fuel cycle. But the volumes will be smaller and the levels of radioactivity less compared to a uranium cycle with comparable actinide recycling and transmutation. A large part of the long-term radioactivity, however, comes from two radioactive substances which not are fission products, protactinium-231 (Pa-231) and thorium-229 (Th-229). Both are aggressive alpha emitters with very long half-lives of 32,760 years and 7340 years, respectively. With 20 times the half-life, this requires storage times of some 650,000 and 15,000 years, respectively (Standring *et al.*, 2008).

Molten lead is envisaged as the target and coolant liquid. This does, however, require a temperature of 700°C, a temperature level at which lead is highly corrosive. Lead–bismuth thus seems to be more realistic, but the multitude of reactions taking place, both spallation and fission, make lead–bismuth highly radioactive. Both radioactive gases (xenon and krypton) and a lot of radioactive volatiles as cesium and iodine are produced, many of them with long half-lives. The levels of radioactivity present become very high indeed, complicating maintenance and making extra shielding necessary (Kara *et al.*, 2008). Any reactor with bismuth as a coolant will also produce the highly radioactive polonium-210 (Po-210). Other very problematic isotopes produced with the envisaged accelerator technology are Po-209 and Po-208 with half-lives of 102 and 2.9 years, respectively (Kara *et al.*, 2008, p. 66). Thus the radioactivity present when decommissioning thorium reactors becomes a major challenge, as do the waste volumes.

A permanent solution to climate change

> Looking at the extensive problems caused by emissions of climate gases, it is clear that it is urgent to develop this new form of energy.
>
> What is more important today than to give the world safe and CO_2-free energy?
>
> (Omtvedt, 2006)

Energy production from large numbers of thorium reactors worldwide has continually been placed within the context of climate change, and presented as a permanent global solution to these problems. Electricity generated from thorium power plants certainly is virtually CO_2-free, at least when considered as a closed system. In the real world, however, all systems are open. Major development of thorium power plants does not take place in isolation. On the contrary, as a major economic endeavour, it will be tightly integrated with other societal sectors and activities that not are CO_2-free, and their CO_2 emissions may increase when subject to major growth. Accepting this idealistic precondition for the time being, there is still a critical issue of time. Things take time, often more time than we like to think, especially when new, complicated technological systems leave the closed system of drawing boards and become implemented in the real open world.

As regards time, it is worth emphasising the time needed for the most relevant European prototype to work. The MYRRHA project in Belgium was started in 1997, and is now expected to be in operation towards 2020, that is more than 20 years to put only this one prototype in operation. In addition to an accelerator, it contains a subcritical reactor core, with a conventional MOX fuel with about 35 per cent plutonium however. The accelerator will only have an effect of 1.5 MW, and the thermal power of the reactor will be some 60 MW, 4 per cent of the envisaged full-size reactor. Coolant and proton target will be lead–bismuth and not pure molten lead, even when we are looking as far ahead as 2020 (Kara *et al.*, 2008).

The Norwegian plans have proposed to build the first prototype accelerator and thorium reactor in Norway. It would at least take 10 years for detailed planning and construction, thereafter at least 5 years for trial operation and verification. Planning and construction of a number of thorium power plants complete with new fuel fabrication and reprocessing facilities would take at least another 15 years, so the first reactors will most probably not be in operation for another 30 years. Completion of a large number of new reactors and power plants globally will take at least 40 years. Most optimistically, then, we are talking about the year 2050 before any major electricity production could take place. The plans have envisaged some 20,000 new reactors. The total current global number of electricity-producing uranium reactors is 445, which has taken the world almost 60 years to achieve.

But 2050 is too late, anyhow. The major reductions of CO_2 emissions, at least 80 per cent globally, must be made before then. Internationally there is today almost total agreement on this time schedule. If thorium power is ever realised, it will not contribute to solving the climate change problem. We simply do not have the time to wait for such large new technological adventures. By then we need to have developed totally different solutions that have already proven their viability (Wise, 2005).

Concluding words: technological idealism

I want to use *technological idealism* as the term pinpointing how we can understand the basis for the research claims about ADS thorium power. As emphasised, these are claims that have come to dominate the Norwegian discourse on energy and climate for some years, a discourse which in this chapter has been subject to a critical analysis. The discourse is real, and the claims it is founded on are just as real, but that does not imply that they necessarily are realistic or even true. Claims like these can be part of reality, but still be false. I have elaborated on the unjustified simplifications and sheer falseness of many of the claims, not only in a technological sense but in some cases also in a pure physical sense. As a reminder, I consider the following claims to have such characteristics:

- The Norwegian thorium resources are proven reserves.
- The economic value of these reserves are a thousand times the value of the Norwegian oil fund.
- It is almost technically impossible to produce nuclear weapons with this technology.
- There are no particularly critical issues connected to the production of U-233; it is only a highly valuable fuel.
- A melt-down due to operator errors is not possible.
- There are no particular accident likelihoods in a thorium power plant consisting of *in situ* online reactor, accelerator, reprocessing and fuel fabrication.
- Waste from a thorium reactor only needs to be stored for 700 years.
- Looking at the extensive problems caused by emissions of climate gases, it is clear that it is urgent to develop this new form of energy.

The claims can be understood as a result of undue idealism. The *idealism* term in 'technological idealism' is understood here as a position within theory of science, basically as a contrasting position to realism. *Absolute idealism* implies that there is no reality beyond, or independent of, our ideas about it (Morgan, 2007).

The main contributors to the thorium power concept and claims are not found in technological sciences, but in basic university natural sciences, such as particle physics and nuclear chemistry. Contributions from particle physicists have been particularly prominent. To the extent that there is a foundation in technological sciences, it is limited to the particle accelerator technology. Thus the original idea and concept were developed by the Nobel Prize laureate *Carlo Rubbia*, a particle physicist at the international particle physics research centre – CERN (Rubbia, 1996; Rubbia *et al.*, 1997). The Norwegian discourse initiator is also employed at CERN. In the last decades, particle physics has developed into a scientific field where idealistic stands are protruding. Purely theoretical works have become common, works where their relation to a reality is largely considered a non-issue. Even in the experimental works their relation to reality remains unclear, and there is a continuous uncertainty within the confined closed systems to what extent findings and results are actually caused by the initiating ideas and the interconnected laboratory equipment.

Technological idealism and *closed systems science* may be understood as major reasons for the fundamental seriousness of contemporary environmental problems and ecological crisis (Høyer and Næss, 2008). Such idealism may also entail *technological oversell*, where the ripeness and benefits of technologies are over-communicated, and the economical, social and environmental problems are under-communicated. It may look innocent but still have dire consequences, as on several occasions lately has been demonstrated within the field of bio- and genetic technologies. In our Norwegian case, it did cause an unfounded thorium Klondike to take place, and two new private companies to be established. It is yet to be seen whether this particular technological adventure will end in a prototype with a cost of at least 1 billion euros.

References

Archer, M., Bhaskar, R., Collier, A., Lawson, T. and Norrie, A. (eds.) (2007). *Critical Realism. Essential Readings*. London: Routledge.

Bhaskar, R. (1979). *The Possibility of Naturalism*. London: Routledge.

Bhaskar, R. (1991). *Philosophy and the Idea of Freedom*. Oxford, UK: Blackwell.

Bhaskar, R. (2008). *A Realist Theory of Science*. London: Routledge.

Brøgger, A. and Høyer, K.G. (1986). [*Det radioactive Norge etter Tsjernobyl*]. The Radioactive Norway after Chernobyl. Oslo: Samlaget Publ.

Buch-Hansen, H. and Nielsen, P. (2005). [*Kritisk realisme.*] *Critical Realism*. Roskilde University, Denmark: Roskilde Universitetsforlag.

Dahlgren, S. (2008). [*Thorium i Buskerud, Telemark og Vestfold fylker.*] *Thorium in the counties Buskerud, Telemark and Vestfold, Norway*. Report. Tønsberg, Norway: The Regional Geologist.

Danermark, B., Ekström, M., Jakobsen, L. and Karlsson, J. Ch. (1997). *Att förklara samhället*. Lund, Sweden: Studentlitteratur.

GEO (2007). [En stor Ressurs – men har vi forekomst?] A large resource – but do we have proven reserves? *GEO*, April. [Norges Geologiske Undersøkelser.] Norway Geological Survey.

Hartwig, M. (ed.) (2007). *Dictionary of Critical Realism*. London: Routledge.

Høyer, K.G., Gundersen, H., Poleszynski, D. and Reinton, P.O. (1977). [*Spillet om atomkraften*.] *The Play of Nuclear Power*. Oslo: Pax Publ.

Høyer, K.G. and Næss, P. (2008). Interdisciplinarity, ecology and scientific theory: the case of sustainable urban development. *Journal of Critical Realism*, **7** (2), 179–208.

Lovins, A. (1977). *Soft Energy Paths: Toward a Durable Peace*. Cambridge, MA: Ballinger.

Lillestøl, E. (2006a). [Norsk grunnstoff kan løse kraftkrisen.] Norwegian metal can solve the energy Crisis. Interview with Professor Egil Lillestøl in *Apollon*. University of Oslo, 4 October 2006.

Lillestøl, E. (2006b). Quote from a lecture at University of Oslo in a discussion with this author.

Lillestøl, E. (2006c). [Den snille atomalderen?] The kind nuclear age? Interview with Professor Egil.

Lillestøl, E. (2006d). Thorium-based nuclear energy. Interview with Professor Egil Lillestøl, in *DIVA international*, Switzerland (www.divainternational.ch).

Lillestøl in *Under dusken*. University of Bergen, 6 November 2006.

Martiniussen, E. (2007). [*Thorium som kjernebrensel*] *Thorium as Nuclear Fuel*. ZERO-notat. Oslo: ZERO.

Megon (1973). [*Thorium-fremstilling i Norge*] *Thorium-production in Norway*. Oslo: Megon Ltd.

Morgan, J. (2007). Idealism. In Hartwig, M. (ed.). *Dictionary of Critical Realism*. London: Routledge.

OECD/NEA (2001). *Trends in Nuclear Fuel Cycle*. Paris: OECD/NEA.

Omtvedt, J. P. (2006). [Thorium, et nytt energieventyr.] Thorium, a new energy adventure. Interview with Professor Jon Petter Omtvedt, *Aftenposten*, 30 November 2006.

Omtvedt, J. P. (2009). [Kjernekraft fra thorium.] Nuclear power from thorium. Lecture by Professor Jon.

Petter Omtvedt, *MILEN-seminar*, University of Oslo, 26 May 2009.

Perrow, C. (1984). *Normal Accidents: Living with High Risk Technologies*. Princeton, NJ: Princeton University Press.

Perrow, C. (1999). *Normal Accidents: Living with High Risk Technologies*. Princeton, NJ: Princeton University Press. New edn.

Perrow, C. (2007). *The Next Catastrophe. Reducing Our Vulnerabilites to Natural, Industrial, and Terrorist Disasters*. Princeton, NJ: Princeton University Press.

Rubbia, C. (1996). *The Energy Amplifier. A Description for the Non-specialist* [online] .

Rubbia, C. *et al.* (1997). CERN-group conceptual design of a fast neutron operated high power energy amplifier, I: Accelerator driven systems: Energy generation and transmutation of nuclear waste. Status report. Vienna: IAEA-TECDOC-985.

Samseth, J. (2007). [Bærekraftig kjernekraft.] Sustainable nuclear energy. ppt lecture, 9 March 2007 [online].

Standeren P.Ø. (2007). Managing director of Thorium Norway Ltd, quoted in the Norwegian publication GEO, April: [Norges Geologiske Undersøkelser.] Norway Geological Survey.

Standring, W.J.F., Hassfjell, C. and Seyersted, M. (2008). *Environmental Impacts from and Regulation of Potential Thorium-based Industry in Norway* (in Norwegian). Strålevernrapport 2008:10. Østerås, Oslo: Norwegian Radiation Protection Authority.

Thorium Report Committee (2008). *Thorium as an Energy Source – Opportunities for Norway*. Report from the Norwegian Thorium Report Committee. Oslo: Research Council Norway.

US Geological Survey (2007). Mineral commodity summaries. January, 2007 [online].
WISE (2005). Nuclear power: no solution to climate change. WISE report, February, 2005. World Information Service on Energy [online].
WNA (2003). *Accelerator-driven Nuclear Energy*. World Nuclear Association [online].

11 Food crises and global warming

Critical realism and the need to re-institutionalize science

Hugh Lacey and Maria Inês Lacey

Activities in many and various domains of human life not only contribute causally to, but also experience harmful impact from, global warming, deriving from the build up of greenhouse gases in the atmosphere, and the climate changes that it is bringing about. Combating global warming, therefore, requires efforts in all these domains to eliminate its causes and to reverse its harmful impact. Agriculture is one of these domains (see p. 188).

Currently predominant agricultural practices are a major source of the build up of greenhouse gases in the atmosphere; and the harmful impact of global warming on agriculture and the food supply is significant. Many features of the food crisis of 2008, and the continued threat of food insecurity facing countless millions of poor people throughout the world, for example, are inseparable from global warming and the related climate changes. Combating the agricultural causes of global warming, then, should be accompanied by efforts to eliminate the threat of food insecurity. One proposal (perhaps the only serious one) currently being made for dealing permanently with the fundamental causes of food insecurity – a system of agricultural production that is based on working everywhere towards local 'food sovereignty' – if implemented on a large scale, would also bring about significant reductions in the emission of greenhouse gases (see p. 190). Global warming and the threat of continuing food insecurity can be combated together. Agricultural policies and practices that credibly promise to eliminate threats of recurring food crises – informed by the appropriate kind of scientific research – can be crucial components of the package of proposals needed to deal with global warming.

In the argument that follows, the links between agricultural practices, food crises and global warming are described and their fundamental causes located in the prevailing capitalist–market agricultural system. This is the basis for an explanatory critique of this system (see p. 188).[1] Then, the proposals to bring about food sovereignty are drawn upon in order to rebut efforts to dull the force of the explanatory critique, which are made by those who claim that there is no viable alternative system (see p. 190). Scientific research is needed to explore the credibility of the proposal that the practices aiming to bring about food sovereignty can provide a viable alternative. Conducting the relevant kind of research, however, requires space for the use of currently marginalized methodologies and a role for broader democratic input into the priorities and character of scientific research. It requires that science be re-institutionalized (see p. 197).

Explaining the food crisis of 2008

The most severe worldwide food crisis in recent decades occurred during the first half of 2008, following the sudden rise of food prices to record-breaking levels. It was marked by increased hunger and starvation in impoverished sectors of the world as large numbers of people became unable to provide adequate food for themselves and their families and, in several countries, it provoked serious social unrest ('food riots').[2]

Why did this food crisis come about? Conventional wisdom, provided by the newspapers[3] (and mainstream science/food policy publications), attributes it to the conjoined impact of four factors:

(i) Sudden and large increases in the price of petroleum.
(ii) New policies encouraging the development of agrofuels.[4]
(iii) New demands coming from 'rapidly developing' countries such as China and India for food products, and especially for meats, and thus for the crops needed to feed livestock.
(iv) Crop failures due to persistent adverse weather conditions in countries that are large-scale food exporters.

Causal mechanisms

The causal mechanisms that connect these four factors to the food crisis are clear enough. The role of (i), increase in the price of petroleum, in contributing to higher food prices is accounted for by three main mechanisms that are related to the widespread use of petroleum and petroleum-derived products in the distribution and also the production of foodstuffs. First, most food products are marketed through the institutions and mechanisms of agribusiness and other large capitalist bodies, e.g. supermarkets, much of it transported long distances nationally and internationally so that transport costs (as well as profits for the various intermediaries between farm and supermarket) contribute significantly to food costs. Second, much farm production is mechanized, using machines that consume large quantities of gasoline or diesel fuel. Third, 'conventional' farming, and also farming based on growing transgenics (GMOs), is heavily dependent on the use of petrochemicals: fertilizers, herbicides, pesticides, and their use is exacerbated as a consequence of the widespread growing of monocultures.

Regarding the role of (ii), with the rising demand for agrofuels and international policies that foster their use, there arises competition to use farmlands, either to grow foodstuffs (and other traditional agricultural products, e.g. cotton, other fibers and tobacco), or to grow crops for agrofuels. Then, unless new farmlands become available, less land will remain available for growing food crops, resulting in smaller amounts of foodstuffs produced and, in the face of increased demand for food, higher prices (Altieri, 2009; Holt-Giménez and Shattuck, 2009; Rosset, 2009a). The mechanisms of the role of (iii) are similar: raising livestock takes land away from plant food production, and also competes with human beings for consumption of plant products. And, in the case of (iv), weather-induced crop failures in major exporting countries lead to shortages of foodstuffs on the international market and thus increased demand for what is available; hence higher prices (Bradsher, 2008b).

These mechanisms relate the four factors *directly* to increases in the price of foodstuffs, and hence, *in view of the prevailing background conditions*, to the food crisis:

(a) Most people gain access to the food they need and want principally by means of buying it in markets that are responsive to international market fluctuations, and
(b) The *status quo* for a significant part of the world's population is one in which vulnerability to and even the immediate experience of hunger, starvation and malnutrition, and the threat of further food crises, are ever-present actualities.

Although with the onset of the financial crisis in the latter part of 2008, and the fall in price of petroleum, prices of food commodities have fallen and the severity of the food crisis has abated, *these two conditions remain in place*. For millions of poor people the condition of food insecurity and constant vulnerability to hunger and malnourishment was only exacerbated, not created, by the food crisis.[5]

In order to understand the 2008 food crisis, we need to answer, not only 'Why did *this* food crisis come about?' and 'Why did the four factors become salient at the same time?' but also 'Why do most people gain access to the food they need and want principally by means of buying it in markets that are responsive to international market fluctuations?' and 'Why do threats of further crises remain?' Factor (i) is no longer operative at the time of writing this chapter (August 2009) and, as stated above, the crisis has abated – and the relevance of (iv) varies with place and time. Nevertheless, while the two conditions (a) and (b) remain in place, the threat of further crises cannot be ignored, a threat that is not assuaged by the continuance of factors (ii) and (iii) and great uncertainties about (iv); and so the need remains urgent to create a system of food production and distribution that is not vulnerable to such crises.

Systemic roots of the causal mechanisms

The causal mechanisms linking factors (i)–(iv) to the 2008 food crisis and to the threat of further food crises have systemic roots, in the system of contemporary agricultural productive and distributive practices. These practices are capital-intensive, for the most part controlled by large agribusiness corporations, industrial, dependent on petrochemical inputs and on technoscientific innovations, e.g. hybrid plants and transgenics, which tend to be implemented by way of planting monocultures. Furthermore, they are integrated into the international market system (regulated by such institutions as WTO and IMF), in which economic growth *per se* is considered to be essential to development, and the foremost aims are to generate profit, to consolidate and expand the control of agribusiness over as many dimensions of agricultural production and distribution as possible, and to satisfy in 'developing' countries consumerist desires and habits comparable to those taken for granted in the 'developed' countries.

Obviously, factors (i)–(iii) represent fluctuations within the international capitalist-market system; and (iv) is connected with another systemically based factor, viz. that agribusiness interests usually lead to the large-scale growing of monocultures. Adverse weather conditions are more likely to cause crop failures, and on a greater scale, when crops are grown in monocultures, especially varieties that require conditions (e.g. availability of a plentiful water supply, or perhaps a predictable temperature range) that are especially vulnerable to climate change (Pittock, 2005; Kaiser and Drennen, 1993). In addition, the severity of the crisis has been related by some observers to additional fluctuations within the system: increased speculation on food commodities aiming for short-term profit, deregulation of markets, and emphasis (backed by government subsidies) on growing crops for export (Rosset, 2009a).

The integration of the production and distribution of foodstuffs into this capitalist-market system ensures that food prices – and hence the availability of food for vulnerable people – will be responsive to market fluctuations and the interests of profit. Food will be considered a commodity like any other commodity (Altieri, 2009). Beneficiaries of this system promise that the proper functioning of the market would enable everyone to gain access to sufficient food. Nevertheless, that promise has never been fulfilled, despite the fact that currently food sufficient to feed everyone is produced and could be available to feed everyone alive today, if there were appropriate mechanisms of distribution.[6] The capitalist market does not provide such mechanisms, since its workings are subordinate to the interest of profit, and so food will not be made available on this market at prices that poor people can afford to pay, unless it is profitable to do so. Thus, the capitalist-market system cannot be counted on when unfavorable market conditions emerge. Consequently, access to food (food security) will not be considered a sovereign right of people. Then, the means of production and distribution of food that might enhance the possibility of food self-reliance will not be developed, and poor people will remain in a permanent state of vulnerability to hunger and starvation. Hence the threat of further crises! The system is the fundamental source of the persistence of the threat of worsened hunger and malnutrition (see p. 191).

Connections between the food crisis and global warming

The food crisis is inseparable from global warming and the climate change that it brings about.[7]

In the first place, global warming is among the causes of the food crisis. With respect to (iv): the adverse weather conditions experienced in many regions, especially the extremes of drought, high rainfall, increased numbers of hurricanes, etc., which lead to crop failures, are probably part of the climate changes caused by global warming (Pittock, 2005, p. 16; Battisti and Naylor, 2009), and they put pressure on the world's useable water supplies (Altieri, 2009; Pengue, 2009; Gommes, 1993).[8] In addition, it is anticipated that the increased temperatures will lead to the death of many forests from heat stress and that many crops will not yield well in their current locations (Pittock, 2005, pp. 45, 108–9, 119) and yields are likely to fall in 'developing' countries (Pittock, 2005, pp. 122, 270; Rosenzweig and Parry, 1993). With respect to (ii): although the new emphasis on agrofuels is part of the response to develop alternative energy resources stimulated by the rising price of petroleum, it also is seen as responding to global warming. Supposedly agrofuels are more 'ecologically friendly', less polluting in use and less generative of the greenhouse gases that are the principal cause of global warming.

Secondly, because of the mechanisms connected with the role of factor (i) (p. 184), the system of agricultural production contributes to the emission of greenhouse gases that generate and sustain global warming and, as factors (ii) and (iii) become more pronounced, is likely to do so on a larger scale.[9] On the one hand, increases in demand for agrofuels and for food products (especially meat) lead to destroying forests, typically by burning them, thereby both adding carbon dioxide to the atmosphere, and eliminating trees that absorb it. On the other hand, despite the claim that agrofuels are 'environmentally friendly', evidence has been put forward suggesting that, to the contrary, growing crops for agrofuels actually increases greenhouse gases in the atmosphere, in part because the process of converting crops (especially in the case of maize) into methanol requires the use of very large quantities of energy that (in many cases) is likely to be derived from fossil fuels (Fagione *et al.*, 2008; Rosenthal, 2008a; Scharlemann and Laurance, 2008; Searchinger *et al.*, 2008; NYT 2008c, 2009). In addition, there is evidence that suggests that growing these crops undermines environmental and social sustainability (Jacobson, 2009) and weakens food security (Altieri, 2009). Also, some of the new crops used for producing agrofuels are dangerously invasive species (Rosenthal, 2008c), and higher temperatures and increased carbon dioxide content of the atmosphere also foster the growth of certain weeds (Christopher, 2008), both of which lead to increased use of pesticides, thereby magnifying the effects of the mechanisms involved in the role of (i).

Since it exacerbates global warming and, in turn, global warming contributes causally to the persisting threat of food crises, the current agricultural system has effects that contribute to undermine its own sustainability and to reinforce its inability to ensure food security for the world's poor. Furthermore, it is plausible to project that efforts to maintain and extend the system will involve additional contributions to global warming. But that may not happen, for it should not be ruled out summarily that technoscientific innovations might make a great difference to what is possible within the system, perhaps even mitigating some of the harmful effects that have been discussed. The current trajectory of technoscientific innovation in agriculture, however, increasingly dominated by agribusiness and exemplified by the widespread use of transgenics, is not promising in this respect, for it entrenches all the mechanisms that relate factor (i) to the food crisis (§1.1) (Lacey, 2005a, Part 2; Altieri, 2009; Rosset, 2009a) and thus to global warming.

An explanatory critique

The current agricultural system causally contributes simultaneously to exacerbate global warming and to maintain food insecurity for many people and nations – and shows no promise of major change in this regard (see p. 192). In line with the critical realist theme of 'explanatory critique',[10] therefore, a negative evaluation should be drawn of the agricultural system – unless alternative modes of agricultural production cannot produce enough to feed and nourish the world's population, or unless they would do comparable or even greater damage in the domain of global warming and its environmental accompaniments. Critical realism alerts us to consider seriously that there may be such alternatives and to engage in appropriate research for the sake of identifying them and investigating their prospects. The real, CR maintains, is not identifiable with the actual (or the dominant trajectories of actual structures and practices), but also includes the possible (including hitherto non-actualized possibilities). As part of explanatory critique, a positive evaluation (again, *ceteris paribus*) should be drawn of practices aiming to actualize alternative modes of agricultural production and distribution, which offer the credible promise of being sufficiently productive to feed everyone everywhere and without the adverse consequences of the prevailing system of production. The power of the critique of the agricultural system to motivate action for change will depend on a positive assessment of the potential of relevant alternative agricultural practices.

Alternative system of agricultural production

Organized social movements throughout the world are proposing an alternative system of agricultural production with the aim of ensuring food security for everyone, and they are engaged in implementing whatever aspects of it that their resources will permit. They do not propose a single alternative to 'conventional' and transgenic forms of agriculture, but rather a variegated array of farming practices – organic, subsistence, biodynamic, agroecological, ecologically sustainable, permaculture, the 'system of rice intensification' (Broad, 2008), and others adapted for use in urban settings (e.g. Royte, 2009) – and the deployment of appropriate combinations and variations of them. The system would be constituted by a multiplicity of complementary locally-specific combinations and variations, each adaptable to its social-ecological environment, that simultaneously are (a) highly productive of nutritious foodstuffs, environmentally sustainable and protective of biodiversity, (b) more in tune with and strengthening of communities of rural people and the variations of their aspirations with place and culture, (c) able to play an integral role in producing the food necessary to feed the world's growing population, and (d) particularly well suited to ensure that rural populations in 'developing' countries are well fed and nourished, so that current patterns of hunger could be abolished.

It is not only social movements of the poor who are attempting to construct an alternative system.[11] Throughout the world today, among many different groups of people, the values of food security, sustainability, healthy foods and local productivity have taken on high ethical salience, and a great variety of efforts are under way to introduce practices that emphasize organic foods (thus rejecting the use of chemical pollutants), attempts to construct new relations between producers and consumers, new forms of marketing goods, questioning of eating habits that require foods the production of which undermines environmental sustainability, decentralized (and urban garden) production. All these efforts fall among the variegated array of practices that may be bringing the proposed new system into being. Although more research is needed before a definitive appraisal of its potential can be made, the prospects appear to be promising.

Such a system would not be vulnerable in the same way as the prevailing one either to market fluctuations or to weather-induced hazards, and it would provide the foundation for local self-reliance in food. It would permit *food sovereignty*, 'the right of peoples and sovereign states to democratically determine their own agricultural and food policies',[12] and it embodies the proposal that food sovereignty is the best means of ensuring *food security*, 'a situation that exists when all people, at all times, have physical, social and economic access to sufficient, safe and nutritious food that meets their dietary needs and food preferences for an active and healthy life' (IAASTD, 2008, executive summary, p. 8).

The conditions and governmental policies needed for the development and maintenance of such a system, and the central role of family and co-operative framing in it, have been elaborated in detail by the social movements – the international alliance of organizations of peasant and family farmers, farm workers, indigenous peoples, landless peasants, and rural women and youth (Rosset, 2009a) – that are part of *Via Campesina* (Via Campesina–Brazil, 2008; Rosset, 2009a). Agroecology is accorded a central role by *Via Campesina* among the various practices that would make up the alternative agricultural system. It is a form of farming that aims to develop and maintain agroecosystems that enable there to be a satisfactory balance of the four desiderata: productivity, sustainability (ecological integrity and preservation of biodiversity), social health, and strengthening of local people's agency (Altieri, 1995). 'Agroecology' also refers to a program of scientific research, whose aim is to investigate agroecosystems with respect to how they fare in the light of the four desiderata, with a view to discovering in all locales the conditions under which they may or may not be actualized in appropriate balance.[13]

The alternative agricultural system and reduction in greenhouse gas emissions

Furthermore, in this alternative system, with its local focus, transportation costs, and the use of petroleum that they imply, would be greatly reduced. Petrochemical inputs would be minimized because the careful design of agroecosystems – rich in biodiversity, that emphasizes growing mixed kinds and varieties of crops (with appropriate rotations), and running farms that produce a multiplicity of products – eliminates much of the need for (artificial, petroleum-derived) fertilizers, herbicides and pesticides. Moreover, it would involve producing food under conditions in which the enhancement of sustainability (respect for nature and maintaining ecosystems, preservation of biodiversity) and social health are more highly rated values that profit or economic growth; then the production of agrofuels (when undertaken) would not be at the expense of food production.

The alternative system, therefore, if it were developed, would find a ready place in the multiplicity of alternative practices that would have to be strengthened throughout numerous domains of human life and activity, if global warming is to be combated in a serious way. Already there is compelling evidence that agroecology, and other forms of farming listed on p. 195, successfully meet the food and nutrition needs of many small farming communities throughout the world, whose needs are not addressed by the prevailing system (for examples, see Altieri, 1995; Pimbert, 2009). Creating conditions and resources to enable these practices to expand should be a matter of urgency, both because of their demonstrated success in serving poor communities, and because only by doing so can evidence be obtained about the potential of the alternative system that would enable a definitive judgment to be made. Here the relevant research cannot be separated from engaging in the practices and recording their outcomes (Lacey, 2002, 2005a, Ch. 11).

Attempting to dull the force of the explanatory critique: 'There is no alternative system of agricultural production' – and rebuttal of the attempts

As stated above, the power of the explanatory critique of the agricultural system to motivate action for change will depend on the positive assessment of the potential of relevant alternative agricultural practices. But the critique, as it stands, does suffice to underline the urgency of conducting research (and providing the necessary resources and conditions for it) to test the potential of a promising alternative.

Beneficiaries of the current agricultural system often respond that no further research is needed, for *the matter is already settled*: outside the trajectory of the current capitalist–market system based on technoscientific innovation that contributes to economic growth, there really is no alternative system of agricultural production that can meet the food and nutrition needs of the world's growing population. Certainly, they say, there is no available scientific evidence to support that claim – even if, in some cases, the alternatives meet the needs of small farming communities (or fill a special niche, e.g. for organic foods, not satisfied by the predominant system), this does not extrapolate to meeting the food needs of large urban populations.

It is important to be clear about what is at issue here. Food security for all, a system that will enable all to be fed and nourished, is the fundamental aim of those advocating the alternative. Although the prevailing system currently produces enough food to feed everyone, it does not have mechanisms to ensure that everyone is fed.[14] High productivity, by itself, is not sufficient to ensure food security. This is recognized by the alternative proposals, which aim to integrate productive and distributive mechanisms at the local level. The claim made against them is that they lack the productive capacity to meet the world's food needs as the population continues to grow, especially as it is more and more concentrated in large urban settings. It is true that *now* the alternative system does not have the productive capacity to feed the world's current population; after all it has not had the conditions to develop sufficiently for its potential to be assessed. The proponents of the prevailing system focus on productive capacity. They claim that only by strengthening current trajectories of agricultural innovation in this system (e.g. with the increased use of transgenics) can adequate food be produced to meet the expected demand for food in the future (Lacey, 2005a, Ch. 10). Is it a settled matter that there is compelling evidence that this is so?

Is the capitalist-market system the fundamental cause of persisting food insecurity?

One party maintains that the current system (and it's current trajectories) cannot bring about food security, the other that the alternative system lacks the needed productive capacity. The latter party, reflecting 'conventional wisdom', also challenges the diagnosis (p. 188) that the capitalist-market system, *as distinct from some of its historically contingent features*, is one of the fundamental causes of vulnerability to food crises and of global warming. If there really are no viable possibilities outside of this system (discussed in Lacey, 2002, 2005, Ch. 11), then the system cannot be the fundamental cause of food insecurity. For the proponents of the system, factors (i)–(iv) – together with the condition (b): 'The *status quo* for a significant part of the world's population is one in which vulnerability to and even the immediate experience of hunger, starvation and malnutrition, and the threat of further food crises, are ever-present actualities' (p. 185) [15] – suffice to explain the 2008 crisis. Moreover, they may explain its severity and the persistence of (b) by a further factor:

(v) Protectionist policies enacted by many countries inhibit competitiveness, with the effect that food production is kept below what it could be and large quantities of food are withheld from the international market, and so they artificially generate further scarcity.

Among the proponents, there is some variation of opinion about the significance of the causal contribution of the various factors. For some, (ii) is the most serious factor: a time of rising costs of production and distribution of foodstuffs is not the time to put further pressure on food costs by taking away agricultural land for the sake of marketing non-food products (NYT, 2008a–d). Others tend to say that, given market mechanisms and the success of development programs in China and India, higher food prices are now here to stay – but they have been distorted upwards by misfortune (iv) and misguided policies (v). The 'solution' for them is a new equilibrium to be worked out in the play of the free market – so, to avoid starvation for large numbers among their populations, impoverished countries will have to find new ways to enter more effectively into the free market so that people will have the money to buy food at the higher prices, and restrictions (e.g. to preserve forests and other matters connected with reducing global warming) will have to be set aside.

Until now, however, condition (b) has persisted within the capitalist-market system, and given condition (a), that most people must buy the food they consume at prices subject to market fluctuations (p. 185), it is difficult to see how such a new (yet to be established) equilibrium would solve the problem. The only alternative to (a) seems to be the provision of much greater aid to impoverished people and nations. Perhaps aid aimed to strengthen local productive capacity (as distinct from just food aid) would make a difference to market conditions. Historically, aid has prevented catastrophe at some times of great calamity. But it has not provided long-term redress to the vulnerability of many people to hunger. It cannot be expected to so, for food aid – in addition to the fact that it can easily generate dependence and reinforce the corruption of the powerful in poor nations – cannot provide a permanent solution, since it is subject to the whims and changing interests of the rich countries.

Despite the persistence of (b) coexisting with ample production of food, the proponents of the prevailing system continue to focus on productive capacity without addressing how the mechanism of production may affect food security issues. When they affirm that there is no agricultural alternative, they are confident that the productivity of their system will increase, because it utilizes ongoing technoscientific innovation in farming. Furthermore, they maintain, these innovations pose no significant risks (when properly regulated) to health and the environment, and they even promise to reverse some of the environmental damage caused by current 'conventional' methods of farming that derive from excessive dependence on petrochemicals. Hence, e.g. the use of transgenics, an exemplary technoscientific innovation, has spread throughout the world and become important in the agricultural policies of many countries, *accompanied by the legitimating claims of 'no alternatives' and 'no serious risks'*. Reflection on the case of transgenics raises general issues pertinent to the kind of research needed to inform practices designed to redress the problems of global warming and to produce a more sustainable and less vulnerable world.

Alternatives and risks – and scientific research

Questions of alternatives and risks are, of course, matters for scientific investigation (Lacey, 2005a, Part 2). As pointed out above (§3.1), more research is needed before the definitive appraisal of the potential of the 'food sovereignty' alternative system aiming to satisfy the food and nutrition needs of everyone can be ascertained. So too more research is needed on the potential of the methods based on technoscientific innovations to be sufficiently productive, at the same time sustainable and free from serious risks, and whether the means of distribution accompanying the productive innovations are adequate to ensure that everyone everywhere can be properly fed. What is the appropriate research to engage in for the sake of reaching sound judgments on these matters, and what methodologies need to be deployed?

Sound agricultural policy should be informed by scientific research that attempts to provide empirically well-grounded answers to the following question about '*the range of agricultural alternatives*': what agricultural methods – 'conventional', transgenic *and* the variegated array of methods listed on pp. 188–9 – and in what combinations and with what locally specific variations, *could* be *sustainable*, relatively free from risks (including those connected with greenhouse gas emissions), and *sufficiently productive*, when accompanied by viable distribution methods, *to meet the food and nutrition needs of the whole world's population* in the foreseeable future?

Methodological considerations

Unless appropriate research is conducted responding to the range-of-alternatives question, the matter of whether or not there is a viable alternative system cannot be considered scientifically settled. What methodologies need to be adopted in order to engage in such appropriate research? Here it is important to keep in mind that seeds used in farming are simultaneously many kinds of things:[16] (a) Biological entities: under appropriate conditions they will grow into mature plants from which (e.g.) grain will be harvested. (b) Constituents of various ecological systems. (c) Entities that have themselves been developed and produced in the course of human practices. (d) Objects of human knowledge and empirical investigation. All these need to be taken into account when investigating risks and the potential of alternatives. In addition, in accordance with another theme of critical realism – that ontology is prior to methodology, that methodology must be appropriate to the kind of object being investigated – deliberations about methodological issues need to enter into the argument.

Decontextualized/reductionist – D/R – methodologies

Sufficiently far-reaching methodological deliberations usually do not take place in mainstream science, however. In it, science tends to be identified with technoscience, research conducted with the horizon of technoscientific innovation in view, and often conducted specifically for the sake of generating such innovation.[17] The methodologies of technoscience deploy a mode of understanding that focuses on underlying molecular structures of phenomena, their physicochemical mechanisms (interactions and processes), mathematical laws and quantifiable properties, and that (consequently) enables discovery of the possibilities for exercising technological control – and, in so doing, they *decontextualize* the phenomena by ignoring their ecological, human and social contexts, and any possibilities that they may gain from being in these contexts and from their relationship to human experience and values, and (in the case of biological and human phenomena) *reduce* them to underlying physicochemical mechanisms.[18]

No phenomena can be *fully* understood without some use of decontextualized/reductionist methodologies (*D/R methodologies*). But, if only they are used, *some phenomena* cannot be adequately understood – including:

- Risks: especially long-term ecological and social risks of technoscientific innovation (Lacey, 2005a, Ch. 9) – and not just risks, but harm already caused by technoscientific innovation under the socio-economic conditions of their implementation, such as that manifested in global warming.
- The causal networks in which problems facing the poor (such as vulnerability to food crises) are located (Lacey, 2005a, Ch. 8).
- Alternative practices (e.g. agroecology) that are not primarily based on using technoscientific innovations as, e.g. practices involving the use of transgenics are (Lacey, 2005a, Ch. 10).
- Phenomena that cannot be reduced to their underlying physicochemical mechanisms: e.g. biological organisms and their developmental stages, ecological systems, human intentional action, and social structures.

To investigate *these four kinds of phenomena*, methodologies that do not decontextualize or reduce, and that are marginalized in mainstream science, must be used.

Risks

Concerning risks, it is not sufficient to consider only *direct risks* to human health and the environment connected with chemical, biochemical and physical mechanisms, that can be quantified and their probabilities estimated (and which can, to a significant extent, be well investigated using only D/R methodologies). *Indirect risks* also need to be considered, i.e., risks that arise because of socioeconomic mechanisms, e.g. *in the case of the widespread use of transgenics*, long-term environmental risks that arise because most transgenics are not only biological objects, open to genomic and molecular biological investigation for example, but also commodities, entangled in issues of intellectual property rights; or risks to social arrangements that arise from the actual context of their use, including risks of undermining alternative forms of farming, and (hence) risks occasioned because extensively using transgenics serves to bring the world's food supply increasingly under the control of a few corporations and so more vulnerable to market contingencies.

Indirect risks, since they cannot be separated from ecological and social context, cannot be investigated adequately only using D/R methodologies. The methodologies appropriate for generating technoscientific innovations are not by themselves adequate for investigating the risks that may be occasioned by the social implementation of the innovations. In mainstream science, given its tendency to identify science with investigation conducted with D/R methodologies, this tends to mean that indirect risks are investigated only when paying attention to them cannot be avoided, and then only in fragmentary, haphazard, sporadic, *ad hoc*, *post hoc*, easily manipulated, opportunistic ways. Mainstream science, for the most part, investigates phenomena only insofar as they can be investigated under D/R methodologies; it effectively subordinates ontology to methodology, since it cannot identify the possibilities open to phenomena that cannot be grasped under these methodologies and (by not raising the question of how to investigate them) effectively takes for granted that they do not exist.

Alternatives

Just as the mainstream marginalizes relevant research on indirect risks by declaring it (since not conducted utilizing D/R methodologies) to be 'not really scientific', so it also questions the scientific credentials of the research that is needed to inform the alternative forms of farming listed above (pp. 188–9) – and, therefore, needed so to be able to reach an empirically-informed judgment about the range-of-alternatives question.

Consider agroecology.[19] In agroecological investigation, the seed is considered as component of an agroecosystem that is investigated in terms of how well it fares in light of the desiderata: productivity, sustainability (ecological integrity and preservation of biodiversity), social health, and strengthening of local's peoples agency (Altieri, 1995), with a view to discovering the conditions under which they may or may not be actualized in appropriate balance. Context is essential; the role and potential of the seed in an agroecosystem cannot be reduced to what can be grasped from attending only to its underlying (genomic and molecular) structures and mechanisms and their physicochemical interactions with other (decontextualized) components of the agroecosystem. The results of molecular biology may inform agroecology in many ways, but molecular biology simply lacks the conceptual resources to deal adequately with the agroecosystem.

Research in agroecology is essentially inter- and multidisciplinary, drawing not only on the mainstream biological sciences, but also on (at least) ecology, sociology, economics, and political science. More, it draws upon indigenous and local knowledge and traditional practices, with which it often manifests continuity. It needs to utilize the farming, observational skills and knowledge of the farmers themselves, who characteristically have a more complete knowledge of the ecosystems that they work in than formally trained scientists do, and also of their histories and of the practices that can be sustained and that maintain biodiversity. Moreover, since they are the ones whose values and cultures are to be strengthened by agroecological practices, agroecological research cannot be conducted without their committed participation. In agroecological research, there is not a clear line between the researcher and the farming practitioner, and between formally trained scientists and the bearers of traditional knowledge. This adds credibility to the scientific credentials of agroecological research. This claim may appear odd, but only where science tends to be reduced to technoscience, and its methodologies to those that explore the underlying mechanisms and laws of phenomena in dissociation from their place in agroecosystems.

Science

Science should be thought of as systematic empirical inquiry, responsive to the ideal of objectivity (Lacey, 2005a, Chs 1, 2) – while recognizing inevitable uncertainties in investigations on, e.g. risks and alternatives (Lacey, 2005b) – conducted using whatever methodologies are appropriate for gaining understanding of the objects being investigated. Then, technoscience and, more generally, research conducted under D/R methodologies, is just one, albeit an important and indispensable approach to science.

Then, indigenous knowledge – and also knowledge gained from, e.g. agroecological, feminist, deep ecological and other perspectives – need not stand opposed to scientific knowledge, and only investigation on a case-by-case basis can establish whether or not its epistemic credentials (and also those that use only D/R methodologies) are deficient for dealing with particular objects of investigation. Traditional knowledge practices, provided that they are subject to empirical constraint (not necessarily constraint from data obtained in the laboratory, but also from 'the test of practice', the exercise of practical 'know how', and 'the test of time'),[20] may reasonably be incorporated under the category of 'science' – noting that, when science is thought of as including a pluralism of methodologies (not only D/R ones, but also those that do not dissociate from context), there is no threat of reducing traditional knowledge-gaining practices to those that exclusively utilize the D/R approach[21] and, in a patronizing way, granting them the status of 'science', provided that they meet the strictures of research conducted within this approach.

Methodological pluralism

Unless science is thought of in this expanded way, permitting methodological pluralism, the range-of-alternatives question cannot adequately be addressed scientifically, for D/R methodologies can deal adequately neither with risks nor alternatives. Then, any claim, made without utilizing the appropriate pluralism of methodologies, that there are no alternative forms of agricultural production (and no serious risks) (see §4), would be simply dogmatic – reflecting either the empirically uninvestigated hypothesis that all phenomena can be grasped with the categories available when decontextualized methodologies are used, or the equally empirically uninvestigated hypothesis that the resources of the D/R approaches (and consequent technoscientific innovations) are able to inform all viable practices.

The range-of-alternatives question remains central. Sound food policies need to be informed by empirical research pertaining to it. Moreover, given the causal links of current predominant agricultural practices with global warming, as well as their systemic links with the vulnerability of poor people to further food crises, there is urgency about investigating the productive potential of the alternative system. And the relevant research needs to be multi- and interdisciplinary, making use of an appropriate plurality of methodologies, and integrated with developments of knowledge that informs traditional and indigenous practices. But currently institutionalized science practically identifies scientific research with that conducted under D/R methodologies and, within it, research priorities are chosen in the light of the strictures of these methodologies and the interest in technoscientific innovation that would contribute to economic growth within the prevailing capitalist–market system. Thus, currently institutionalized science is unable to inform policy makers reliably on relevant matters of risks and alternatives. It tends to dismiss proposals, like those made for the priority of food sovereignty, not on the basis of the results of relevant research framed by the range-of-alternatives question; rather, since they are not amenable to being investigated using only its favored methodologies, they do not even become candidates considered for investigation.[22] Hence, if the proponents of food sovereignty are right, currently institutionalized science – by prioritizing research that might lead to technoscientific innovation, rather than that framed by the range-of-alternatives question – contributes to maintain the state of food insecurity for many poor people.

It also emphasizes looking to find technoscientific innovations that might contribute to alleviating the problem of global warming, rather than exploring that potential of alternatives like agroecology, which lie outside of the trajectory of the capitalist–market system and are not based on technoscientific innovation, and which depend on changes in relations of human beings with nature and with one another (without excluding an essential role for technoscientific innovation).

Re-institutionalized science

In order that the range–of–alternatives question may be addressed, therefore, science needs to be re-institutionalized. The re-institutionalized science would have broad democratic participation and oversight, in order to redirect the uses of scientific knowledge and the priorities of research, to make use of important methodologies that are currently marginalized, and to create space where researchers can *begin with* the aspirations, assessments of needs, and practices of the social movements (like *Via Campesina*), and involve their participation in an integral way. Then, the forms that science takes, and the kinds of questions it addresses, could be determined in collaboration with the social movements and reflect their values and experiences. The proposal is not intended to deny space for research aiming for technoscientific innovation, but to create institutional forms in which there can be democratic deliberation – involving the participation of representatives of all who experience the impact of technoscientific innovation, and who have proposals for dealing with the world's serious problems – about appropriate priorities for research and allocations of resources.[23] Above all, it is to enable resources to become available for research that could test the potential of alternatives and inform their conduct; and it would insist that the range–of–alternatives question be thoroughly investigated, concerning risks (including causal connection to global warming) and alternatives, before technoscientific innovations be socially introduced.

Concluding remarks

The current predominant system of agricultural production causally contributes significantly, not only to the vulnerability of many poor people and nations to food crises, but also to the quantity of greenhouse gases present in the atmosphere and thus to global warming and the climate changes that accompany it. Hence, the explanatory critique, that this system should be negatively valued. In general, however, the power of an explanatory critique to motivate action for change, depends on identifying proposed courses of action for eliminating the object criticized (in this case, for transforming the system of agricultural production) that are positively valued. Concerning issues connected with global warming, the motivation for change has been difficult to generate; even though the authoritative reports put out by IPCC (International Panel on Climate Change) offer compelling models of anticipated dire climate change, they have not instigated much governmental action to deal with its causes. After noting this fact, Revkin (2009) quotes a leading climate scientist: 'For IPCC, this means providing guidance that will minimize climate impacts and maximize investments in a prosperous and sustainable future'.

To motivate change, no matter how dire the outlook of continuing the current trajectory, a positively attractive course of action needs to be at hand. Hence, the extended discussion of the possibility of an alternative system of agricultural production, the forms it might take, and the agents who would develop it. If global warming is to be contained, the 'investments' will have to be in many areas. Agriculture is one of them and, connected with it, the practices aiming for food sovereignty should be leading contenders for investment, and also the forms of scientific investigation that can be expected to inform these practices. Certainly, it should not be presumed, prior to appropriate scientific investigation, that only forms of scientific research that can inform technoscientific innovation are relevant. Hence, the need to re-institutionalize science, so that proposals, e.g. concerning the importance of food sovereignty, that simultaneously promise to address the vulnerability of poor people to food crises and to contribute to redressing global warming, can be investigated fully and (as appropriate) implemented.

Notes

1 Explanatory critique is a central theme of critical realism (see Note 10). In making the argument, other themes of CR are also drawn upon, *viz*., that the real is not identifiable with the actual but includes also the possible, and that methodology should be subordinate to ontology and appropriate to the kind of object being investigated.

2 FAO (2009b); Rosset (2009a) For newspaper reports see, e.g., Bradsher (2008a) and Lacey, M. (2008). Lacey reports food riots in Guinea, Mauritania, Mexico, Morocco, Senegal, Uzbekistan and Yemen; Rosset (2009b) in Bangladesh, Brazil, Burkina Faso, Cameroon, Côte d'Ivoire, Egypt, India, Indonesia, Mozambique, Pakistan, Myanmar, Panama, the Philippines, Russia, Senegal, and Somalia.

3 See, e.g., Krugman (2008). USAID (2008). Re. (ii): see Martin (2008); re. (iv): Bradsher (2008b).

4 'Agrofuels' – fuels produced from agricultural products, often called 'biofuels'.

5 'International food prices have come down from their 2008 peaks, but are higher than they were in 2006 and likely to remain volatile. In many developing countries, the cost of staple foods remains stubbornly high. The financial crisis is straining the ability of the poor to cope. Easing the burden of high food prices at the outset was critical – but more needs to be done. Efforts now need to focus on building farmers' resilience to future shocks and improving food security over the long term' (FAO, 2009a).

The financial crisis led to the fall in prices of food commodities. But, in other ways (e.g. through increasing unemployment) it created additional difficulties for poor people to buy food.

'The double whammy of high food prices and the economic meltdown has pushed more than 100 million people into poverty and hunger. Although international prices have come down from their record highs in 2008, they have yet to drop to their levels before the food crisis, and the risk of volatility continues. Average food prices in May 2009 were about 24 percent higher than they were in 2006. And, in many developing countries, the cost of basic food staples is stubbornly high. Unemployment and reduced wages, remittances and government services – by-products of the economic slump – threaten to add to the woes of the world's poorest people, who already spend between 60 and 80 percent of their income on food' (FAO, 2009a).

6 This claim has been widely documented; see, e.g., Boucher (1999), and it is not contested in the discussions of the 2008 food crisis.

7 The following websites provide up to date references to the literature on this topic: Science and Development Network, http://www.scidev.net and Food First, http://www.foodfirst.org.

8 'Water sources will become more variable, droughts and floods will stress agricultural systems, some coastal food-producing areas will be inundated by the seas, and food production will fall in some places in the interior. Developing economies and the poorest of the poor likely will be hardest hit' (Nelson, 2009).

'Water scarcity and the timing of water availability will increasingly constrain production. Climate change will require a new look at water storage to cope with the impacts of more and extreme precipitation, higher intra- and inter-seasonal variations, and increased rates of evapotranspiration in all types of ecosystems. Extreme climate events (floods and droughts) are increasing and expected to increase in frequency and severity and there are likely to be significant consequences in all regions for food and forestry production and food insecurity. There is a serious potential for future conflicts over habitable land and natural resources such as freshwater. Climate change is affecting the distribution of plants, invasive species, pests and disease vectors of many human, animal and the geographic range and incidence of many plant diseases is likely to increase' (IASSTD, 2008, executive summary, p. 15).

9 'Today, agriculture contributes about 14% of annual greenhouse gas emissions, and land use change including forest loss contributes another 19%. The relative

contributions differ dramatically by region. The developing world accounts for about 50% of agricultural missions and 80% of land use change and forestry emissions' (Nelson 2009). 'Agriculture presently contributes about 21–25%, 60%, and 65–80% of total anthropogenic emissions of carbon dioxide, methane and nitrous oxide, respectively . . . Agriculture is also thought to be responsible for over 90% of the ammonia, 50% of the carbon monoxide . . . released into the atmosphere as a result of human activities' (Duxbury and Mosier, 1993, p. 232).

10 The idea of explanatory critique – that a negative evaluation should be drawn (*ceteris paribus*) of the causes of the social acceptance of false beliefs and (*as in this case*) of negatively valued phenomena, and a positive evaluation (also *ceteris paribus*) of courses of action rationally chosen for the sake of removing the causes – has been thoroughly developed in several writings by Roy Bhaskar, most fully in Bhaskar (1986), Ch. 2, Sects. 6, 7. For an overview, analysis, and a lot of references, see Lacey (2007).

11 The goal, food security for all and the minimization of threats of further food crises, is shared by many international (humanitarian, aid and agricultural) agencies, which reject the reliability of the capitalist market for meeting this goal, and point to the need to enhance and protect local food productive capacity in all countries. Some agencies endorse proposals close to those of the social movements (e.g. IAASRD); for others the goal of obtaining food security is not to be at the expense of strengthening modified mechanisms of the international market.

12 Cf., food sovereignty 'as people's right to healthy and culturally appropriate food produced through ecologically sound and sustainable methods, and their right to define their own food and agricultural systems. . .. It requires an immediate moratorium and eventual rollback of agrofuels . . . It relies on agroecological approaches to production and protects the farmer's right to seed, land, water, and fair markets. Food sovereignty requires the democratization of our food systems – their spaces and places – in favor of the poor' (Holt-Giménez and Shattuck, 2009).

Rosset (2009a) summarizes 'Food sovereignty policies to address the global food price crisis' as involving the following demands:

- Protect domestic food markets against both dumping (artificially low prices) and artificially high prices driven by speculation and volatility in global markets.
- Return to improved versions of supply management policies at the national level and improved international commodity agreements at a global level.
- Recovery of the productive capacity of peasant and family farm sectors, via floor prices, improved marketing boards, public-sector budgets, and genuine agrarian reform.
- Rebuild improved versions of public sector and/or farmer-owned inventories, elimination of transnationals and the domestic private sector as the principal owners of national food stocks.
- Controls against hoarding, speculating, and forced export of needed foodstuffs.
- An immediate moratorium on agrofuels.
- The technological transformation of farming systems, based on agroecology, to break the link between food and petroleum prices, and to conserve and restore the productive capacity of farmlands.

13 See Lacey (2005a), Part 2, for discussion of the evidence for and extensive documentation of the productive potential of agroecology as an agricultural practice, and a defense of the sound scientific credentials of agroecological research.

14 Condition (b) is considered a historically contingent feature of the system – or, if food security cannot be ensured within the system and so cannot be ensured at all, then (b) would represent just the tragic fact that scarcity is part of the human condition. Of course, those who experience food insecurity have every motive to demand that the

potential of all alternatives be investigated urgently and with provision of adequate resources, that practices that demonstrably meet their needs be expanded, and not be stopped short by easy projections from the current state of affairs or the interests of the beneficiaries of the prevailing system. It is also important to keep in mind that highly motivated organized action to bring about an alternative can be a crucial causal factor in realizing an alternative; the potential of alternatives cannot properly be appraised independently of this factor and so it cannot be 'read off' from current predominant trajectories (Lacey, 2002, 2005a, Ch. 11).

15 Seeds are examples of 'laminated systems' (Bhaskar and Danermark, 2006).

16 The nature of technoscience and its characteristic methodological features are discussed in more detail in Lacey (2008b). A fuller account is not needed here, for what matters, so far as the present argument is concerned, is only that research in technoscience involves the use of D/R methodologies. Note the characterization of 'science' below (§4.4), which does not limit 'scientific' research to that conducted under D/R methodologies, and which recognizes the scientific status of the methodologies used, for example, in agroecology (Lacey, 2005a, Part 1).

17 These methodologies, the reasons for their being virtually exclusively deployed in modern science, and the possibility of a pluralism of methodologies (not all of which are reducible to those of the decontextualized approach) under which objective scientific knowledge may be gained, are discussed fully in Lacey (1999, Chs 6–10; 2005a, Part 1).

18 For further details, see Lacey (2005a), Chs 5, 10.

19 Remember that traditional knowledge informed the selection practices that bequeathed us the seeds that are indispensable for growing all crops today, and without which transgenics would be impossible.

20 There is a growing literature showing the richness, variability, versatility, sensitivity to sustainability issues, and empirical soundness (that is not undermined by being reflective of the interests and values of particular cultural groups) of much traditional and indigenous knowledge (e.g. Pimbert, 2009; Santos, 2007). As 'science' is being used here, it can incorporate all these forms of knowledge, while retaining their specific features and not forcing them into a shape that supposedly fits all scientific research; and they become indispensable resources for addressing – scientifically – the range-of-alternatives question. The authors cited here prefer to talk of these forms of knowledge, not as 'scientific', but as 'other knowledges' ('decolonialized knowledges'), terminology that they intend to have relativist connotations. Whether or not these other forms of knowledge are to be called 'scientific' is not very important; the important things are their sound empirical credentials, and that having these credentials does not depend on using D/R methodologies. The connoted relativism is unnecessary (and unfounded). What is present here is not knowledge relative to particular cultures, but approaches to investigation that are properly reflective of the character or aspects of the object being investigated – aspects that may be considered important because culturally specific values are held. This does not make the knowledge, as distinct from its significance, relative to these cultural values.

21 That D/R methodologies are used almost exclusively in modern science is linked either with commitment to materialist metaphysics, that all possibilities can be grasped with the categories deployed in (current or still to be developed) D/R methodologies (Lacey 2009), or by making certain assumption about technological progress, e.g. that all the great problems of the world, including dealing with harm caused by technoscientific innovations themselves, can be resolved by technoscientific innovation, and typically only in that way (Lacey 2005a, Ch. 1). Either way, assumptions are involved that could not be confirmed by research exclusively conducted using D/R methodologies. However, they are deep in the 'common sense' of modern science, and so taken for granted, as well as being powerfully reinforced by current forms of funding for research that emphasize that research should lead to contributions to economic growth, that

their empirical status is seldom thought about. This makes it difficult for the sound claims of alternatives to gain a hearing within mainstream science.

22 One might put it: the claim is, in the name of human rights, for a niche for research that does not reflect market relations.

23 This is a general claim. How it would be worked out in areas, other than agriculture, is beyond the scope of this chapter (see Lacey, 2008a).

References

Altieri, M.A. (1955). *Agroecology: The Science of Sustainable Development*. Boulder, CO: Westview.

Altieri, M.A. (2009). The ecological impacts of large-scale agrofuel monoculture production systems in the Americas. *Bulletin of Science, Technology and Society*, **29**, 236–244.

Battisti, D. S. and Naylor, R. L. (2009). Historical warnings of future food insecurity with unprecedented seasonal heat. *Science*, **323**, 240–244.

Bhaskar, R. (1986). *Scientific Realism and Human Emancipation*. London: Verso.

Bhaskar, R. and Danermark, B. (2006). Metatheory, interdisciplinarity and disability research: a critical realist perspective. *Scandinavian Journal of Disability Research*, **8**(4), 278–97.

Boucher, D.H. (ed.) (1999). *The Paradox of Plenty: Hunger in a Bountiful World*. Oakland, CA: Food First Books.

Bradsher, K. (2008a). High rice cost creating fears of Asia unrest. *The New York Times*, March 29. <http://www.nytimes.com/2008/03/29/business/worldbusiness/29rice.html?.ex=1207454400&en=cd90b06bce3a6924&ei=5070&emc=eta1>.

Bradsher, K. (2008b). A drought in Australia, a global shortage of rice. *The New York Times*, April 17. <http://.www.nytimes.com/2008/04/17/business/worldbusiness/17warm.html?ex=1209096000&en=dbac6be3b5db5a78&ei=5070&emc=eta1>.

Broad, W. J. (2008). Food revolution that starts with rice'. *The New York Times*, June 17. <http://www.nytimes.com/2008/06/17/science/17rice.html?scp=1&sq=&st=nyt>.

Christopher, T. (2008). Can weeds help solve the climate crisis? *The New York Times*, June 29. <http://www.nytimes.com/2008/06/29/magazine/29weeds-t.html?scp=1&sq=&st=nyt>.

Duxbury, J.M. and Mosier, A.R. (1993). Status and issues concerning agricultural emissions of greenhouse gases. In Kaiser and Drennen (1993), pp. 229–258.

Food and Agricultural Organization of the United Nations (FAO). (2009a). <http://www.fao.org/isfp/about/en/>.

FAO. (2009b). Secretariat contribution to defining the objectives and possible decisions of the World Summit on Food Security on 16, 17 and 18 November 2009. <http://www.fao.org/fileadmin/user_upload/newsroom/docs/Secretariat_Contribution_for_Summit%20.pdf>.

Fargione, J., Hill, J., Tilman, D., Polasky, S. and Hawthorne, P. (2008). Land clearing and the biofuel carbon debt. *Science*, **319**, 1235–1238.

Gommes, R. (1993). Current climate and population constraints on agriculture. In Kaiser and Drennen (1993), pp. 67–86.

Holt-Giménez, E. and Shattuck, A. (2009). The agrofuels transition: restructuring places and spaces in the global food system. *Bulletin of Science, Technology and Society*, **29**, 180–88.

IAASTD: International Assessment of Agricultural, Science and Technology for Development (IAASTD) (2008). *Agriculture at a Crossroads*, http://www.agassessment.

org/. Executive Summary <http://www.agassessment.org/docs/SR_Exec_Sum_280508_English.htm>.

Jacobson, M.Z. (2009). Review of solutions to global warming, air pollution, and energy security. *Energy And Environmental Science*, **2**, 148–173.

Kaiser, H.M. and Drennen, T. E. (eds.) (1993). *Agricultural Dimensions of Global Climate Change*. Delray Beach, FL: St. Lucie Press.

Krugman, P. (2008). Grains gone wild. *The New York Times*, April 7, <http://www.nytimes.com/2008/04/07/opinion/07krugman.html>.

Lacey, H. (2002). Explanatory critique and emancipatory movements. *Journal of Critical Realism*, **1**, 7–31.

Lacey, H. (2005a). *Values and Objectivity in Science: Current Controversy about Transgenic Crops*. Lanham, MD: Lexington Books

Lacey, H. (2005b). On the interplay of the cognitive and the social in scientific practices. *Philosophy of Science*, **72**, 977–988.

Lacey, H. (2007). Explanatory critique. In Hartwig, M. (ed.). *Dictionary of Critical Realism*, pp. 196–201. London: Routledge.

Lacey, H. (2008a). Crescimento econômico, meio-ambiente e sustentabilidade social: a responsabilidade dos cientistas e aquestão dos transgênicos. In Dupas, G. (ed.), *Meio-ambiente e Cresimento Econômico: Tensões estruturais*, pp. 91–130. São Paulo: Editora UNESP.

Lacey, H. (2008b). Ciência, respeito à natureza e bem-estar humano. *Scientiae Studia*, **8**, 297–327.

Lacy, H. (2009). The interplay of scientific activity, worldviews and value outlooks. *Science and Education*, **18**, 839–860.

Lacey, M. (2008). Across globe, empty bellies bring rising anger. *The New York Times*, April 18. <http://www.nytimes.com/2008/04/18/world/americas/18food.html?ex=1209182400&en=3e8a729773aaf389&ei=5070&emc=eta1>.

Martin, A. (2008). Fuel choices, food crises and finger-pointing. *The New York Times*, April 15. <http://www.nytimes.com/2008/04/15/business/worldbusiness/15food.html?ex=1208923200&en=9c715f2f2c497b48&ei=5070&emc=eta1>.

Nelson, G.C. (2009). Agriculture and climate change: an agenda for negotiation in Copenhagen – Overview'. IFPRI: International Food Policy Research Institute, <http://www.ifpri.org/publication/agriculture-and-climatechange>.

NYT (2008a). The world food crisis. *The New York Times*, Editorial, April 10.

NYT (2008b). Rethinking ethanol. *The New York Tines*, Editorial, May 11.

NYT (2008c). Honesty about ethanol. *The New York Times*, Editorial, November 17.

NYT (2009a). Getting ethanol right. *The New York Times*, Editorial, May 23.

Pengue, W.A. (2009). Agrofuels and agrifoods: counting the externalities at the major crossroads of the 21st century. *Bulletin of Science, Technology and Society*, **29**,167–179.

Pimbert, M. (2009). *Towards Food Sovereignty: Reclaiming Autonomous Food Systems*. London: International Institute for Environment and Development. Available at <http://www.iied.org/pubs/display.php?o=G02493&n=1&l=3&k=Towards%20Food%20Sovereignty%20Reclaiming%20autonomous%20food%20systems>.

Pittock, A. B. (2005). *Climate Change: Turning up the Heat*. Melbourne: CSIRO Publishing.

Revkin, A.C. (2009). Nobel halo fades fast for panel on climate change. *The New York Times, Science Times*, pp. 1, 4, August 4.

Rosenthal, E. (2008a). Europe, cutting biofuel subsidies, redirects aid to stress greenest options. *The New York Times*, January 22, <http://www.nytimes.com/2008/01/22/business/worldbusiness/22biofuels.html?scp=1&sq=%91Europe%2C+cutting+biofuel+subsidies%2C+redirects+aid+to+stress+greenest+options%92&st=nyt>.

Rosenthal, E. (2008b).Biofuels deemed a greenhouse effect. *The New York Times*, February 8.

Rosenthal, E. (2008c). New trends in biofuels have new risks. *The New York Times*, May 21.

Rosenzweig, C. and Parry, M. L. (1993). Potential impacts of climate change on world food supply: a summary of a recent international study. In Kaiser and Drennen (1993), pp. 87–116.

Rosset, P. (2009a). Agrofuels, food sovereignty, and the contemporary food crisis. *Bulletin of Science, Technology and Society*, **29**, 189–193.

Rosset, P. (2009b). Food sovereignty in Latin America: confronting the 'new crisis'. *NACLA Report on the Americas*, May–June, 16–22.

Royte, E. (2009). Street farmer. *The New York Times Magazine*, July 5. <http://www.nytimes.com/2009/07/05/magazine/05allen-t.html?scp=1&sq=Street+farmer&st=nyt>.

Santos, Boaventura de S. (2007). *Another Knowledge is Possible: Beyond Northern Epistemologies*. London: Verso.

Scharlemann, J.P.W. and Laurance, W.F. (2008). How green are agrofuels? *Science*, **319**. 43–44.

Searchinger, T., Heimlich, R. Houghton, R.A., Dong, F., Elobeid, A. and Fabiosa, J. (2008). Use of US cropland for agrofuel increases greenhouse gases through emissions from land-use change. *Science*, **319**, 1238–1240.

US AID: United States Agency for International Development (US AID) (2008). US AID responds to global food crisis. <http://www.usaid.gov/our_work/humanitarian_assistance/foodcrisis/>.

Via Campesina – Brasil (2008). O Problema dos Alimentos: a agricultura camponesa é a solução! Pamphlet. Brasilia: Via Campesina.

12 Towards a dialectics of knowledge and care in the global system

Jenneth Parker

Introduction

Climate change challenges us to develop new forms of ethics that can provide common frames to underpin global agreements and can help to motivate and guide us in the major social changes that are needed in response.[1] In this chapter I will consider how the resources of a dialectically conceived critical realist interdisciplinary ontology (DCRI) might combine with some aspects of communitarian, feminist and ecofeminist ethics of care to provide some ways forward. I am interested in how DCRI understandings of the common human condition (concrete universality) in ecological community interplay with understandings of the concrete singularity of the embodied subject. How might this illuminate the nature of our duties to care, including the relationship between local and more global duties? The latter is of great importance in considering effective moral bases for people's participation in both agitating for, and helping to fulfil, climate agreements.

I will argue here that knowledge and care are in a dynamic relationship that should be more explicitly developed: caring effectively is the prime motivation for the increasing development of interdisciplinarity – but we also need interdisciplinary knowledge in order to help us to care effectively. My discussion will be linked to the ethical orientation of people's social movements, understood as (potentially) including contributions from all sectors of society. I see our self-organisation as the main hope for an effective response to the global system crisis of which climate change is a key part. I propose that the conceptual resources outlined here can help to lay the basis for a new ecological humanism. This can keep alive the project of human emancipation through self-understanding (Bhaskar, 1986), the development of moral agency that can both promote flourishing and challenge power that blights that flourishing. This informed moral agency can develop mitigation, adaptation and regeneration strategies as effective responses of care to climate change. A disposition to care is not sufficient – effective care requires knowledge.[2]

The embodied subject in biotic community

Fisher and Tronto (1991, p. 40) have expressed the systemic maintenance aspect of care in the following way:

> Berenice Fisher and I defined care as 'a species activity that includes everything that we do to maintain, continue and repair our "world" so that we can live in it as well as possible.' That world includes our bodies, our selves, and our environment, all of which we seek to weave in a complex, life-sustaining web.

Feminism and communitarianism have covered some of the same ground in relation to the ethical subject but feminists have added in the crucial element of consideration of the embodied subject (Held, 2006; Benhabib, 1995). Some feminists have also accused liberal ontology of dishonesty as they have claimed that it is based on denial of the essential requirement for care throughout the human life cycle. It is claimed that liberal disembodied theory has distorted our view of the human condition. Feminists have often stressed the actual material outcomes of liberal theory, for example most of our physical public environments are constructed for unencumbered, able-bodied, male individuals (Wendell, 1996; Rapp and Ginsburg, 2004). From an ecofeminist viewpoint, the disembodied nature of the liberal moral subject has precisely enabled destructive attitudes to the biosphere because it has denied moral significance to selves as embodied – as if embodiment were a kind of contingency.

Feminists have asserted the importance of considering the singularity of particular circumstances and the ways in which contextual care practices are seen as paying attention to real human needs, and as viewing particular relationships as morally significant in human life. Virginia Held (2006) has recently proposed that the relatedness of human beings is the foundation of the ethics of care. Some theorists of care have argued that whereas universalist theories enjoin upon us a duty that we should treat everyone the same, contextual theories recognise the moral value of special, particular relations. Here I aim to explore the implications of a contextual practice of ethics that recognises universalisabilility (but is sceptical about universality),[3] specifically with regard to an ethics that can help us respond to the challenges of climate change.

Feminist ethics has been particularly concerned to question and protest against the way that the abstract subject and ethical universalism lead to the exclusion of moral voices:

> a constant impulse to return to the details of care processes and structures in life is the starting point of care as a theoretical perspective . . . a shift occurs in what counts as 'knowledge' in making philosophical and political judgments. This shift, then, is not only in terms of abstract ideas, but in whose voice should count.
>
> (Tronto 1995, p. 145)

It is an extremely important part of feminist awareness that the concerns of subordinated and marginalised peoples become 'backgrounded', silent and/or taken for granted. The dialectical relationship to knowledge is important here, for example, it has required constant pressure and attention to the facts through research to achieve the recognition of the extent of domestic violence and child abuse (Walby, 1999). In the same way it is necessary both to research and to provide access for marginalised voices in responding to climate change (Curtin, 2005). Many kinds of existing ethical commitments are consistently under-rated, including the positive social contribution made by carers, and unpaid workers of all kinds, that helps to maintain the web of life and social relations. People involved in these activities are carrying out important duties. The nature of this duty is identified by Sabina Lovibond with reference to the Hegelian concept of 'concrete ethics' in the following way:

> Sittlich obligation enjoins the individual to maintain, or recreate, an already existing social practice which, because of his personal contribution to the task of maintaining it, is also the objective expression of his own identity . . . the idea of an obligation to sustain the institutions which embody a shared way of life . . .
>
> (1983, pp. 63–4)

Consideration of community emphasises the aspect of care that focuses on the key ethical importance of maintaining the texture of moral life. In keeping with women's awareness of the range of bodily maintenance activities, care also emphasises the maintenance of patterns of human relationality as the very basis of ethical practice. Communitarians have often assumed that specific forms of human relations such as the heterosexual family are the most important forms. Feminist forms of communitarianism can stress a variety of human forms of relationality, which could be less patriarchal and less ethnocentric than the model of the western family (Frazer and Lacey, 1993).

Ecofeminists propose that the community of moral worth is the biotic community, or the community of life. This approach stresses the sharing of life processes, or textures, for example the nutrient cycle, or more simply, the seasons. Maria Mies and Vandana Shiva (1993) describes women subsistence farmers' environmental actions as expressions of their situated subjectivity as part of biotic community. Val Plumwood (1993) explores in depth the ways in which dominating forms of subjectivity are constructed in ways that avoid and nullify possible recognition of the biotic community as morally worthy. Ecofeminist recognition of the moral significance of human membership of the biotic community involves a prima facie duty to keep the supportive community in being.

Care as described here is different to justice because care alleges that people are entitled to what they need as part of the moral community. Care theorists centrally reaffirm the moral response to need rather than an abstract assessment of desert: 'People are entitled to what they need because they need it: people are entitled to care because they are part of ongoing relations of care' (Tronto, 1995, p. 146).

However, this also indicates the important duties that are involved in being a member of the community of 'on-going care'– duties that go beyond the bare minimums of justice.

Interdisciplinary ontology and the moral community

Interdisciplinary work aims to bring fields together in a process of mutual re-definition. This harmonises with the ecological or systems approach that it is partly through relations that things reveal themselves. A further point here, related to the notion of 'critique' (see below), is the view that: '. . . things studied in isolation will not have their contradictions adequately exposed to the critique they deserve' (Outhwaite, 1999).

I would also argue that interdisciplinarity can refocus appreciation on previously marginalised aspects of work, pointing to their importance in the new context created by connections. This is in addition to exposing limitations in the scope of limited disciplinary accounts and their resulting weaknesses – as argued in this volume by Petter Næss (Chapter 4).

The interdisciplinary ontology explicated by Bhaskar and Danermark (2007) is explicitly linked to the need for effective responses of care to individual human needs. On their reading, human being represents a complex configuration of physical embodiment, social and relational situatedness, cultural constructions and subjective identities. Bhaskar and Danermark make clear that utilising any one limited theoretical approach on its own will result in ineffective care for the individual. In this way their article exemplifies a dialectics of care and knowledge, but also a dialectics of theory and situated practice:

> Thus following our investigation (and the substantive research results of one of us), we were able to come to the meta-reflection of an attempted real definition of the field of disability studies as an articulated lamination . . .
>
> (Bhaskar, Chapter 1, this volume, p. 7)

This DCR concept of 'lamination' indicates a system that contains relations of emergence and dependence – such as the fact that social and cultural dimensions of this system emerge from, and are dependent on, physical embodiment. The anti-reductionist aspect of DCRI recognises the emergent real powers of the social and cultural and insists that they cannot be reduced to the bio-physical. In relation to the global system, this schema can be represented by nested systems, as in Figure 12.1.

In relation to views of human emancipation, this account of the differentiated holism of human being can help to formulate a new interdisciplinary humanism that can recognise the drama and worth of human struggles for self-development. Different aspects of the human condition are not necessarily harmonious or self-evidently organised into priorities. The human condition can be dilemmatic and in this way we all deserve and need each others' compassion and understanding.

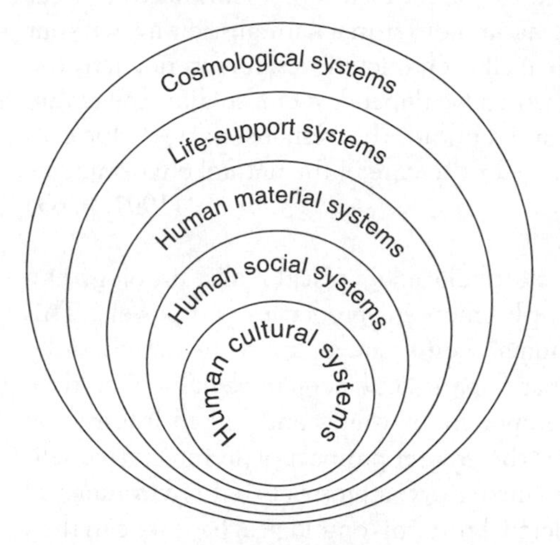

Figure 10.1 Emergence and dependency of systems

We cannot necessarily prioritise the flourishing of this or that aspect, although we may try to arrive at the right relation between different aspects of our being in certain circumstances.

Ecological humanism and flourishing: self-understanding and human emancipation

An interdisciplinary, laminated account of the human condition plus situatedness in ecological space provides the foundations for an ecological humanism. This could be a very important focus, keeping alive the project of human development in a context of global crisis where individualistic survivalism may well test our resources of hope and our capacity for appreciation and recognition of our fellow humans. Crucially, in terms of mitigation of, and adaptation to, climate change, such an ecological humanism can clearly outline the dimensions of human being with which solutions need to engage. Thus it is pointless to attempt to engage and motivate people simply on the basis of rational appreciation of the facts – we will need linked strategies for climate response that appeal to, and work with all the various levels of laminated human being. In the context of developing joined-up knowledge an interdisciplinary ecological humanism can keep alive the commitment to human emancipation through self-understanding as well as through actions that challenge and absent harmful power relations.

Robin Attfield (1991) has developed an account of environmental ethics which is broadly continuous with ethical accounts of human flourishing. In this context he has worked from the principle that faith should be informed by reflection and reason. His approach is one example demonstrating the possibilities of a theistic approach to personal and political action which exhibits an openess to dialogue.[4] Attfield's specification of the metaphysical requirements for an environmental ethic is a prescription that critical realism can fulfil:

> A metaphysics . . . which is suited to our ecological problems needs to treat humans alongside the rest of the natural order in a naturalistic way, without being reductionist about their irreducible characteristics. It must not deny the reality of the natural systems on which we depend, yet must allow the reality of their individual members, and uphold the responsibilities which as individuals and groups people have for the care of the natural environment.
> (1991, p. 63)

Contributions in this volume have articulated a critical realist ontology of a complex, differentiated real world with emergent properties and powers. This ontology provides for the fact that human beings are a part of the natural order (or biotic community) but are also participants in emergent social and cultural processes, one of which is the development of values and linked intentional practices and reflection. This includes the greater prospect of an integrative self-understanding that could be truly emancipatory, a prospect that environmental ethicists to date have not fully considered. From this ontological basis we can then revisit the question of moral community and moral agency that has caused some concern in environmental ethics.

Moral community and moral agency

Ecofeminist accounts of caring for nature could be better formulated in terms of a broadened conception of moral community, rather than attempts to transfer a human ethics of care to nature. Communitarians typically insist upon the importance of shared experiences and public goods (Sandel, 1982) and this is an obvious route to take in making links with environmentalism. However, the communitarian position originates from the perspective of a human sharing of goods and thus presents a narrow form of anthropocentrism. Communitarians do not focus on the sharing of the biotic community in the web of life. However, there seems nothing in principle to prevent the communitarian presumption in favour of the moral significance of shared aspects of life becoming the moral significance of the sharing of life itself. Further, communitarians propose that members of human communities have duties by virtue of their membership and this perspective can equally well apply to membership of the biotic community.

The critical realist ontology of biologically rooted humans with real emergent social and cultural properties and powers can assist here. This ontology can help to clarify a distinction between two senses of membership of moral community. In the first sense the membership is of the community of the moral worth all the members of which exhibit some morally relevant similiarities (Paden, 1994) – the sharing of life. In the second sense membership is of the community of moral agents who share some kind of common understanding and commitment to morality. This distinction leaves open the question of what is due to the widely different members of the community of moral worth. This is the concern of the community of moral agents and is our unavoidable responsibility. Knowledge is essential to explore these questions and involves paying attention to the specific conditions of flourishing of each different class of members. Ethical consistency is important in comparing cases within and across different classes of members – although there may be many different interpretations of what can count as 'the same act' for example.[5]

Cuomo also argues that an uncritical ecological holism that views humans as simply another species is problematic:

> Because holistic perspectives consider humans most commonly as a species, they cannot accommodate inquiries concerning the relationships between the harm humans do to each other and the harm we do to the non-human world.
>
> (Cuomo, 1998, p. 107)

Cuomo points to the necessity of recognising that humans are a species which is capable of ethics and, indeed, in need of ethics in consequence of the degree of power humans possess in the world.

Soper says that she does not believe that to create a layer of 'second-class' moral citizens could be beneficial to animals or living systems (1995, p. 172). However, I argue that moral worth is not a second-rate moral designation as it includes all those beings and entities that are morally worthy but it does not demand the same treatment for all. However it recognises that one thing that is morally relevant about humans is that they are capable of moral agency. In fact we demand moral agency of humans and it is a part of human flourishing that moral agency should flourish. The differences (which may also be on a continuum given research into ethical primate behaviour) are between entities that are moral agents and those that are not.

Curtin raises the question of what exactly is involved in developing 'the capacity to care'. I argue that attempts to discover the facts are a necessary part of care. In caring for other humans there is an assumption of dialogue in that they can let us know what, in their view, constitutes caring for them and we can then take this into account and adjust our behaviour accordingly. In this respect we have a duty to pay attention to their voice as part of a commitment to care. With regard to animals and living systems we have to take responsibility for the construction of an account of what constitutes their flourishing. As we are discovering in the case of climate change, and as contributions in this book testify, we need interdisciplinary knowledge in order to do so.

Clearly, in many cases of wild animals in rich ecologies their flourishing will be assured primarily by humans leaving them alone. However all areas of the planet are inhabited by human communities and their flourishing is also a matter of concern. Many indigenous communities have detailed knowledge of human practices that are conducive to the flourishing of the biotic community as a whole which may be conceptualised as including humans (Bird-David, 1993). This indigenous knowledge is a valuable moral resource of practices embodying values. Some members of the community of the morally considerable may impose the extra duty of care to attempt to discover what our moral duties are in each case. In these cases scientific information will be particularly important. If we do not maintain a fact/value distinction we will be condemned to merely projecting our values in probably inappropriate ways.

The correspondence, or lack of it, between environmental and human political systems requires further attention which it cannot receive here. However, in this context it is important to note that one key environmental proposal is that of bioregionalism which asserts that political structures should more closely correspond to the environmental structures if we are to begin to politically address environmental care. Part of the underlying argument here is that ecological system relations need to be maintained and conserved but human social and political structures can be designed. This is the major reason that membership of the biotic community calls for innovation as well as conservation and recovery of environmentally and socially benign practices. Our challenge as a moral species is to find ways to design human social and economic structures collectively in order to enable the flourishing of the wider moral community and to each other. Progressive social movements for sustainability seek to help us identify and respond to this challenge.

Ontology and advocacy

Charles Taylor has discussed the relationship between ontology and advocacy in the liberal-communitarian debate (1995). Here Taylor is specifically discussing the relationship between the communitarian ontology of the social moral subject, which Taylor glosses as 'holism' and the liberal espousal of the autonomous moral subject, glossed as 'atomist'. In this context he has contributed a thoughtful formulation which I propose to develop further. He has argued that ontology does not determine what we can advocate but it constrains what we can meaningfully advocate.

> Taking an ontological position doesn't amount to advocating something; but at the same time, the ontological does help to define the options it is meaningful to support by advocacy.
>
> (Taylor, 1995, p. 183)

Taylor is thus denying that we can 'read off' values from our assertions of what exists, but he is claiming that nontheless there is an important relationship. In this way Taylor is reinforcing the critical realist claim of the importance of ontology,

> . . . once you opt for holism, extremely important questions remain open at the level of advocacy; at the same time your ontology structures the debate between the alternatives, and forces you to face certain questions. Clarifying the ontological question restructures the debate about advocacy.
>
> (Taylor, 1995, p. 202)

This model can apply to the general relationships between ontology, understood as asserting the reality of features of the human condition, and ethics, seen as principles and practices based upon assertion of these features as morally significant.

Ethical diversity and indeterminacy

Those of a logical turn of mind might ask for a specific model of the constraint that ontology exercises on advocacy – some satisfying counterpart to logical entailment perhaps. However, I believe that Taylor uses the term 'meaningful' advisedly. Taylor's thesis is informed by a Wittgensteinian perspective on meaning which fully allows for its complexity within a community of social relations (1995, p. 96). From this point of view the nature of the constraints on ethics that flow from the multiple features of our ontology will be in need of exposition in each kind of case. There is a very clear linkage in the later Wittgenstein between meaning and action; actions display what an individual believes to follow from a statement and hence enlarge upon the statement. As Paul Johnston puts it,

> . . . understanding means coming to understand the meaning a person's action has for him and this is achieved by examining the context of action, the background against which it has its meaning.
>
> (Johnston, 1989, p. 37)[6]

Humanity is an embodied social species, inhabiting a world in which embodiment has many physical implications. There are constraints upon the positions that we can meaningfully call 'ethical' because of the implications of embodiment and sociality.

What we believe to be the ontology of the human condition constrains what we can reasonably hold to be ethical and this may be changed by science or other knowledge. For example our membership of the class of self-maintaining (living) systems may be reasonably interpreted to be ethically important. Our membership of our human group may also be interpreted as ethically important and in different ways, for different reasons. This example should demonstrate that the nature of the constraint exercised by ontology on advocacy has to be argued through in a multiplicity of ways. This is a very positive aspect as it allows for the recognition of diversity and openness and of their value in enabling ethical development.[7]

Feminist ethicists and others have claimed that ethics can be seen as 'benevolently circular', as moral theory can both explain and critique moral reactions, but moral experience can also cause us to revise our ideas and theories (Dwyer, 1998). Ethics is manifested in human practices, but that moral knowledge from this source can also be reflexively used to critique these practices. This circularity of ethics has implications for moral reformers in dictating that their aim must be to add to or adjust our conceptions of the ethical otherwise they run into great difficulties in losing any basis for the use of the term 'ethical' at all (Midgley, 1984).[8]

Caring particularly

Care theorists argue that contextual approaches can recognise the actuality of our embodiment and constraints that this implies: we all live in a particular place or places and we can all only relate deeply to a limited number of other people and places. This contextual aspect of care approaches to ethics stressed the particular narratives of care which are excluded by an exclusive identification of morality with the universal. Tronto, for example, is concerned to recommend a method of moral enquiry which,

> . . . should develop theories that are capable of being compatible with these particular judgements as with abstract principles derived from emotional experience.
>
> (1993, p. 31)

The validity of caring particularly forms one crucial point of difference between care and justice. Universalistic models of justice do not recognise the specificity of embodiment and the finite range of close human relations. The abstract subject is thus enjured to take the concerns of those on the other side of the world to be equal with those of his or her own children, for example, or be accused of moral failure. On the contrary, care theorists assert the universal (generic) value of particular relations of care. This means that we may have a duty to care, but we do not in all cases have a duty to care equally. Care theorists have recognised the condition of embodiment in a real finite world, within a context of finite human and ecological relations.

Liberalism cannot recognise these elements in its abstract citizens and consequently presents the worst of both worlds. It holds back on substantive criticism of power in the liberal state, promoting a fallacious notion of abstract equality, thus weakening its power for social critique and change. On the other hand it logically imposes abstract moral duties on individuals which fail to consider the actual concrete singularity of human lives (Held, 2006). The pattern of other lives with which we have to do, of places that we inhabit and with which we are connected, forms some kind of guide to our moral duties (Starhawk, 1988). This is inescapably individual, cannot be universalised and provides the material for our individual moral narrative. Caring particularly is of great importance in validating all kinds of local action in support of particular environments and communities (Niezen, 2003). Further these particular relations which themselves are important framework features of the human condition allow for degrees of caring and discovery of value that cannot be achieved at the universal level.

The deep appreciation of the universal values in life, and qualities of life, is only possible through real, specific engagement. This is why it is inevitable that the allegedly morally superior universalised aggregates of value employed in development calculations always tend towards legitimising destruction in the abstract general interest. These calculations obscure the obvious realist point of concrete singularity that we all have to live somewhere specific and need to relate to specific people. Abstract calculation also ignores the obvious point that a planet full of ruined places is a ruined planet – we do not want those who believe in an abstract planet to learn the hard way. From the position of embodied singularity that care perspectives propose, there are good reasons to believe that it is only in these particular relationships that we are fully able to appreciate the value of lives and of places more generally. We have a duty to defend that value when it is threatened,[9] this provides the ground for solidarity with others who are defending the places that they care about particularly. Together these places comprise they planet. In this way caring particularly should undercut the accusation of NIMBYism[10] which is often made by to trivialise these episodes of protection of local and specific value. An accusation of NIMBYism is simply an indication that there is a need for a fuller account of our responsibilities across nested systems from the local to the global.

To take a systems perspective together with an ethics of care is to re-prioritise the embodied social individual as the centre of moral action.[11] An ethics of care asserts the universal moral importance of particular and specific relations in human life. Our responsibilities to care for ourselves include looking after our bodies, our mental health and our moral health. These responsibilities extend out to include close personal others and making time for relationships which seem so vital to human well-being. Further as we are profoundly social beings we cannot fully distinguish between our own well being and that of significant others. However, asserting the personal as the centre of an ethics of care has deeper structural implications when allied to ecological systems perspectives. Ecological perspectives view the body as a natural system in interaction with other systems. Equally discussions of the ethics of food, clothing, heating, medicines and so on all point to the ethical relevance of the interactions between the body and other systems. Ecofeminists and others assert that our particular interactions set up clear responsibilities. Where there is a material interaction there is a morally significant interaction which requires assessment and response. This also applies in terms of social relations, and all those areas where social and material relations interact, particularly production and consumption (Curtin, 2005).

We are increasingly living in a global social political system which sets up relations between producers, consumers and investors and claims that these relations are morally neutral. Globalisation has emphasised the connections across local-local material interactions and consumption and production at a great distance. Taking moral responsibility for these relations at any point is a political act as it fundamentally challenges this ideology of moral neutrality. Moral narratives of caring lives affirm the responsibilities associated with our particular interactions in life. This has been expressed as taking responsibility for our 'ecological footprint', but proponents of care would want to extend this conception to our footprint on the whole of the biotic community, including our impact on human social relations. The acceptance of direct moral responsibility for local-to-local and other global relations represents a major change away from a model of delegation of our international responsibilities to an elite body of statespeople. The acceptance of this moral responsibility is inextricably linked to the growing demands for some form of global democracy where power is more evenly distributed across the globe.

Systems perspectives emphasise the role that individual and local small-scale ethical action can play in changing systems. However, understanding what the whole requires must now be a part of local care. Coming to an agreement as to how this global care should be locally and regionally apportioned is a matter of great urgency and difficulty. Macro problems of environmental change need to be identified and the search for solutions has to begin at all levels of political debate and action. We need a much fuller discussion of responsibilities at local, national, regional and global levels.[12] Furthermore levels of action in the global system need to reinforce one another. In this context one challenge must be to enter into debates about the kinds of decision-making most suitable for each level, the kinds of collective action suited to each level and how they can support each other (Drainville, 2004). A critical realist account of emergent properties, at different 'levels' of the laminated system, and at different scales (e.g. the social), can assist with this research and development programme.

Knowledge and care in response to global system crisis

I would argue that the care response to needs is essential in the crisis of the global system. It is not sufficient, nor perhaps even possible to trace responsibility for particular problems to justify demands for radical redistribution of access to planetary resources – even with the constant recognition that those we wish to help must define their own priorities (Gronemeyer, 1993). Further, knowledge of the causal relations in the current global economic system can indicate that often the best form of help is not substantive aid but the ceasing of extant practices of global domination (Sachs, 1993; Lacey and Lacey, Chapter 11, this volume). This points to the importance of understanding and acknowledging some of the key features of the systemic relations which we currently inhabit with others and developing correspondingly better accounts of causality more suited to a systemic perspective. This will involve a commitment to maintaining the network of relations that support a healthy planetary ecosystem – including responding to key threats such as global warming.

Caring in this wider sense will involve existing knowledge of ways to repair and maintain systems of relations but will also require special attention to discovering the facts. This point may apply to knowledge concerned with the maintenance of human systems of relations as much as it applies to maintenance of ecological systems. We need information from social science, psychology and community studies about conditions that enable or disable human communities and individuals (Turner, 1988). This need for knowledge also emphasises the importance of negotiating with local, tacit and indigenous knowledges in relation to human social and ecological systems. Further, moral resources are to be found in culture and the arts. In addition to religious texts, human narratives form rich resources of moral reflection – whether they be in novels such as *War and Peace*, soaps such as *Eastenders*, Bollywood spectaculars or indigenous creation myths. However, it may be argued that we are in need of new narratives that can help to explore our cultural responses to the relatively new knowledge of our global interconnectedness and the rediscovery of our dependence on planetary life-support systems.

The realist view emphasises that care is essential for embodied human beings and that the autonomous hero is a fantasy built upon the servicing activities of others. This perspective has been strengthened by the professional discussions of care in nursing, but also by work on maternal caring. Some feminist ethicists claim that care is primary to justice because without care we could not exist,

> Care seems to me the most basic moral value. As a practice, empirically described, we can say that without care we cannot have life at all. All human beings require a great deal of care in their early years, and most of us need and want caring relationships throughout our lives.
>
> (Held, 1995, p. 131)

This argument essentially relies upon a realist account of human needs, drawing upon the commonality of the complex human condition to provide a 'thicker' version of the good society, or even, given the threat of climate change, the **possible** society.

Social movements and ethical action

Social movements are seen to cut across traditional class boundaries in interesting ways (Eder, 1993; Touraine, 1981). Political theory based on a model of maximising individual interests supports a limited analysis of oppositional political action in terms of class conflict. Conversely it is argued that social movement actors are capable of, and at least are partly driven by, relatively altruistic values (Cohen and Arato, 1994; Melucci, 1996; Jasper, 1997). In this way social movements place themselves outside of traditional class-based political structures and at a tangent to traditional conceptions of 'the political'. Social movement theory takes seriously the cultural, expressive and subjective dimensions of movement resistance and seeks to explain how these dimensions mesh into and support new forms of collective action (Stammers and Eschle, 2005). This approach helps to set new and important questions about identities of collective actors in relation to ethical action.

One key way in which we can initiate change is to engage in activities which develop capacities that we feel to be valuable (Solomon, 1980). This underlines the vital importance of the ability of social movements to provide and agitate for opportunities for practice of the values they espouse and promulgate. Moreover, in developing an ethic we are necessarily extending the boundaries of any activist community, since an ethic is necessarily an intervention and a moral claim upon others. This is one aspect of the commitment to universality of ethics. However, it also imposes on 'us' the necessity to address others as themselves potential, or actual, moral agents.

Human structures have developed as the result of specific and contingent histories and are organised in ways which reflect entrenched power relations, not according to systems levels. With regard to power and responsibility it is accepted that where there is no power or influence there can be no responsibility. However, we have to consider that it may be that morally we *should* have more influence, if influence is apportioned according to large power differences in society. As universal moral agency is a good in itself, which we should seek to further, we therefore have a duty to ensure that we have that degree of influence which allows us to be a competent moral agent in our situation (Wollstonecraft, 1929). This provides a central moral reason for challenging power (in addition to the capacity of power to blight flourishing) and for seeking to ensure that it is distributed throughout the human social system. This also provides a general moral basis for a commitment to bring about more participatory forms of governance in the world. The recognition of agency implied by care can also re-focus notions of 'activism' on the significance of small-scale, everyday practices in recognition of the importance of activities that support the web of life, in both human communities and the wider biotic community.

Employing an interdisciplinary systems approach has the merit of explaining the necessity for collective ethical action. A key feature which necessitates collective action is that social, political, economic structures and dominant discourses mediate our relations. Ethical movements take action to change these mediating structures and seek new structures which can institutionalise or enable more ethical relations (Stammers, 2009). Taking cultural action seriously involves responsibilities to assess dominant discourses in the communities of which we are a part, to challenge and to attempt to change them where necessary (Frank, Chapter 6, this volume). This involves ideological critique where material interests are linked to the maintenance of certain dominant discourses. Social movements help to provide the oppositional discursive and cultural resources to engage in effective cultural action in social systems. When linked to an understanding of constraining structures, systems perspectives mitigate the ethical responsibility of the individual self. Conversely they also highlight the duties of selves towards changing the collectivities in which they are involved and the duties of joining with others to challenge immoral structures.

Politics and ethics revisited

One key strength of value-driven movements as I have described them is that they recognise and foreground the fact that politics is about ethics. It is a great strength to accept and develop this awareness, which appears as an honest recognition in place of appeals to a supposedly value-free political science. It appears as honest because it involves an abandonment of claims to certainty implicit in the idea of 'scientific socialism' for example, and a concomitant rejection of the power of such claims. It may be that values always were a strong sub-text even in organisations that were ostensibly guided by political theory and that these can now be seen as coming to the fore owing to the falling away of a unifying political analysis. However, these radical political analyses also suppressed their roots in ethics. This theoretical denial of ethics made impossible ethical reflexivity in relation to action, leading to cynicism and manipulation. Most importantly this weakens the challenge to further unethical accumulations and uses of power such as those that laid the foundations of the Stalinist state for example. There is therefore hope that political forms explicitly founded on ethics can welcome (and even maybe 'institutionalise') diverse identities and contextual practices of reflexivity, change and critical renewal (Della Porta, 2005).

The challenge for movements is to maintain and foreground their ethical roots whilst maintaining the capacity to act. As ethics is seen to have a direct relationship to forms of cultural moral reflection, movements' cultural action signals their involvement in moral change and reform. Culturally the assertion of the real dependence of humanity on a healthy planetary ecosystem is inevitably construed as an oppositional political act in a global society that is culturally dominated by monetarist realism.[13] Equally feminist and ecofeminist assertions of the reality of human interdependence are oppositional cultural acts in a global society dominated by a liberal atomist model of the individual, as is this book itself (Table 12.1).

Moral responsibility

Theory can give us some ways to map our duties and can tell us the kinds of considerations which ought to have an influence, but only we ourselves can decide how to respond. In this way ethical duty is open – but unlimited, bearing in mind that care for self should be a part of our considerations. Movements help to provide ways in which we can more easily exercise some of our responsibilities, at local, intermediate and global levels (Hopkins, 2001; Scott-Cato, 2008). This is one reason I stress the research element of movements, often carried out by specific NGOs, but also in the sense of people's action research through action and reflection. We need to research ways that can make our collective good will bear fruit.

It may be asked what reasons an interdisciplinary naturalism can provide for us to follow such guidelines of principle, or to agree even to such broad commitments to the moral significance of key features of the human condition. I have argued that morality cannot provide justification other than practice; there are no intellectual pyrotechnics that can substitute for the moral recognition of the better life. This is why the ecofeminist movement and other progressive ethical movements have to be active; without practice we do not have a morality. Movements can provide arguments and facts to back up their ethical claims, but none will be decisive to the person who really cannot see why they should care. The most that ethical theory can do, and it is a lot, is to provide some kind of reasonably coherent account that helps to increase the conceptual viability of the best moral intuitions that we have developed so far. Ethics can help us to decide for ourselves what is at stake in our decisions, and the kinds of considerations that we should review as part of our decision making process. In this, theory must pay tribute to the ethical insights of practice – but practice can then also be developed through the reflection which theory assists. We also need other forms of cultural reflection, generally under-rated and ignored in moral philosophy based in only the rational aspect of human being. These arguments regarding the justification of morality notwithstanding, If we are to engage in collective ethical action then ethical theory is necessary. To involve others and to work together we need to be able to outline some account of why we are taking particular actions, that appeals to principles for which we can provide support. More understanding and discussion of our ethical aims and differences should help our alliances be more effective.

Table 12.1 Summary of the interdisciplinary naturalist ethic outlined in this chapter

Ontology	*Model of ethics*	*Commitments*
Embodied and vulnerable individual – reality of interdependence between humans and between humans and other forms of life	Commitment to the central ethical significance of these features of the human condition	Maintaining webs of relations – social and biological – look for systemic consequences of acts and practices We should value what we need – but also things in themselves
Systemic dialectical model of relations	Limitless responsibility (no end to processes of care) but also mitigated responsibility (according to capacity in context)	Exploring material relations and their ethical implications
Human community of moral agents: community of the morally considerable the whole biotic community	Moral agency as a moral good in itself	Care for humans as necessarily including empowerment and moral agency Human responsibility to care for biotic community
Knowledge of the world plural, fallible and incomplete	Need to know about objects of care and how to care for them	Commitment to finding out about the world Attention to marginalised voices
Social ethical self formed in community – morally responsive – attention to wider community of life raising issues of human 'biological subjectivity'	Rational and emotional response Ethical responsibility for states of the self – but also mitigated by social resources available	Care for self and others in the context of a community of relations. Commitment to enabling personal, subjective and affective change in conjunction with collective change
Role of culture in forming and critiquing human social ethical self, meanings and responses	Responsibility for participation in and promulgation of discourses, cultural meanings etc.	Challenging power is virtuous practice in economic, political, social and cultural forms such as discourses and representations Supporting the construal of politics as ethical
Human systems are mutable , ecological life-support systems are relatively less so	Attempt to harmonise values of social care with values of maintenance of life support systems	Challenging harmful material practices and the structures which enable them Exploring possible changes to human social systems consistent with the set of ethical commitments
Sexed human individuals mediated through culture – different powers	Significance in contestation	Explore the significance of different biological capacities and needs of the sexes

Working with existing moral resources

On the basis of discussions so far it is necessary to recognise the variety of different approaches to already existing values which can be taken by movements. This is a very important consideration as no moral reform takes place *de novo* but always utilises existing moral resources in society. This represents one of the most important differences in the move from action guided by 'scientific' political theory to political action driven by values. In globalising society, movements can draw on moral resources from a wide variety of global areas. These moral resources are not only cultural, for example explicitly formulated moral and/or spiritual codes, but are also the resources of social structures and practices. Where new moral groupings are seeking to change subjectivity in particular ways they can draw on the experience of other social groupings that also respect similiar principles.

Movements employ appeals to already existing values in society. In this approach moral criticism is directed against practices which allegedly are not respecting these existing social values. This may develop into a deeper critique of structures that shape or constrain practices. Movements also attempt moral reform where the emphasis is on changing existing morality, or extending certain aspects of it, leading to revaluation of existing practices in society, both positive and negative. This approach may use research theories, well attested facts, cultural redescriptions, narratives and other art forms to change moral perspectives and valuations and deploy examples from other cultures. Ethical movements attempt to restructure material, social and cultural life to enable ethical practices by setting up small-scale alternative forms of community and/or developing ways to reform some particularly problematic and immoral material and social relations in society, at local and global levels.

Reducing moral alienation

I argue that the interdisciplinary naturalist perspective outlined provides for an understanding of moral alienation, thus enriching the account of human motivation in ethical movements. This position assumes that no facts about morally considerable beings can be neutral and there is no such thing as a morally neutral fact. Facts will always call forth our moral responses. However, there are many cases in which we do not know how to respond morally. Whilst, as encultured beings, the world presents itself as valued, this is not a continuous seamless picture, but one with gaps, discontinuities and uneven development. Too many gaps and contradictions can lead to severe moral alienation. This is especially true in the case of new technologies and new knowledge which can change key aspects of the framework features of the human condition, providing a situation of moral crisis. Ethical movements are well placed to respond and to debate moral responses and to show how these responses articulate with existing beliefs and commitments. Many technologies raise questions about the kinds of human beings we want to become, and hence deep moral questions about the nature of the good life, which cannot be addressed by hegemonic liberalism.

Further, new technologies and knowledge are developed and applied in contexts of political and economic power, and movements are well placed to challenge the vision of the good life (or the lack of it) which such applications involve. Our relatively recent knowledge of global warming present such a crisis in very stark terms and the responses of liberalism to date are looking dangerously ineffectual. Many are concerned that the weak liberal ethical response will encourage the trend towards 'disaster capitalism' that feeds on crisis (Klein, 2007).

Global citizenship as an aspirational virtue ethic

Alasdair MacIntyre's (1984) virtue ethics have been seen within the communitarian philosophical frame given his stress on the virtues of fulfilling social roles and duties. He has also agreed with feminist ethicists that pure ethical principles crucially require contextual interpretation to produce any substantive guidelines for practice (MacIntyre, 1984). MacIntyre's concept of practices is important in describing a vision of a non-alienated moral existence where, '. . . the practice is not just a means to, but is partially constitutive of the good pursued' (Poole, 1991, p. 147).

MacIntyre claims that the 'goods' of ethical identity must be internal to a 'way of life'. The problem is that MacIntyre's examples of non-alienated ethical practice are drawn from traditional societies and he does not have a vision of how to develop such practices to meet the challenges of contemporary life. However, future-facing movements can appropriate this approach to develop aspirational virtue ethics, such as global citizenship. Critics have claimed that global citizenship is a nonsensical formulation as we do not have a global state to confer citizenship upon us – but they miss the point. Global and local human citizens of the wider biotic and human communities are aspiring to an identity that we are struggling to inhabit, and our embodied practices that we create in attempting this are, in Poole's formulation of MacIntyre, 'partly constitutive of the good pursued'.[14]

In terms of moral reform and development, movements fulfil the social role of exploring contradictions and deepening and changing moral descriptions. However I argue that one cannot deeply evaluate a moral position unless one practices it with others. Consistency with principles itself has to be evaluated through practice as it is through examples that we gain our understanding of meaning and significance. However it is important to also link the role of movements with changing practices, and this is particularly relevant in the case of sustainability movements. Moral alienation is partly about not being able to practice the values that we do have. Movements help us to work out how to take action that is more in accordance with our values and this involves strategic action to change social structures. Last but not least is the question of the moral alienation brought about by the ideological and practical constraints of capitalism. Under liberalism, capitalism is presented as beyond human agency – in fact as a condition of agency. What would it mean for us collectively to take moral responsibility for capitalism and to create economies that supported the flourishing of the biotic community and of humans (Costanza, Chapter 8, this volume)?

Climate change – mitigation, adaptation, regeneration

With regard to the international scale it is the backgrounded facts of the realities of people's lives, in their struggles for survival and maintenance of their environments and livelihoods that must claim attention. This harmonises with the broader environmentalist/sustainability perspectives of critical consumerism; here the emphasis is upon pursuing the facts and bringing to light the hidden relationships and social structures of production and exploitation that underlie commodities. In summary, because we live in a real world it is morally incumbent upon us to find out about it: lack of knowledge reduces our capacity for effective moral agency.

As part of the wider sustainability movement we are involved in attempts to rethink and rework all of our daily practices of engagement with the natural world and with each other. Maintaining an ethic is not possible without attempting to live or express our values in material and cultural practices. We may wish actions and practices to express commitments but it is not always clear how to transform practices so that they do so. In this respect movements are ethically driven research programmes to attempt to discover how, individually and collectively, we can transform our practices to reflect our values. In this we have to be open to the possibility that new knowledge will affect our valuations. Transforming practices requires knowledge of their material and social consequences and of social, political and physical contexts of human existence. In this way we have a responsibility to pursue knowledge in the context of a commitment to care, whilst recognising that this knowledge can be from a wide range of sources – from peer reviewed science, to the narratives and practices of non-literate peoples, and everything in between.[15]

Much of this volume is about how to transform knowledge in reponse to climate change in order to inform mitigation strategies, such as reduction of greenhouse gases. We also need to develop adaptation strategies in response to the degree of climate change that is now inevitable due to existing levels of greenhouse gases in our atmosphere. However, a care perspective also demands that we begin to conceive of a project to enable regeneration of our most vital planetary systems. Research and experience show that, if humans adopt the right caring strategies, informed by knowledge, it can be possible to encourage damaged ecosystems to regenerate (Lui, 2005; Society for Ecological Restoration, 2008). This response strategy is the most holistic of all, as it can increase human welfare and contribute to renewing stable micro-climates, making a contribution to the amelioration of climate change (CARE, 2009). As opposed to an atomistic individualism that views ethics as self-sacrifice, relational critical realist approaches can demonstrate that the commitment to maintain a dynamically related and co-dependent system of life can also be the ground of the free enjoyment of our radical singularity.

Notes

1. Such has been the origin of new concepts like 'Contraction and Convergence' (Meyer, 2000), influential at Kyoto, based on the principle of equal use of atmospheric resources by the world's citizens.
2. See Seddon (2008) for a stunning account of how 'plausible but essentially wrong' managerialist ideas have radically undermined the provision of services in the UK.
3. My approach here is that to hold that an ethic must be universalisable means that the it must be capable of being ethically meaningful to any moral agent, not that all moral agents should treat everyone the same.
4. This is crucial as any reconstruction of politics towards ethical commitment should enable more fruitful alliances with those millions of people who adhere to all the varieties of non-fundamentalist religious faith around the world.
5. See section on ethical diversity on p. 213.
6. This is part of the reason that lack of opportunities to practice our morality prevents us from developing it further.
7. This is a key part of the inspiration that I take from Bhaskar's *Dialectics: the pulse of freedom* 1993.
8. For example, Mary Midgley claims that Nietszche's project to reform the whole of morality became incoherent (Midgley, 1984, pp. 39–41).
9. This is not to say that this duty would or should necessarily be always judged to be primary, simply that it is a perfectly respectable and defensible duty, to be taken seriously along with other, say more global duties.
10. NIMBY stands for 'Not In My Back Yard' and is intended to indicate reprehensible narrow self-interest in protecting what one has, or has access to, against the interests of the common good often supposedly served by destruction.
11. Not the same thing as the whole of moral action.
12. See for example Colin Hines (2008) arguing for nested levels of economy.
13. Some scientists have been bemused by the hostility shown to climate science as they tend to regard science as politically neutral – but the cultural implications of climate science are inescapably explosive.
14. For example, the Lammas Low-Impact community, running a concrete utopian experiment in zero-carbon living (Science Shops Wales); the Transition town movement; the Fair Trade movement.
15. In my 2001 reference, I discussed how DCR could support the recognition of local and indigenous knowledges without falling into the relativism that can be seen in some forms of post-modern and post-colonial approaches.

References

Agger, B. (1992). *Cultural Studies as Critical Theory*. London: Falmer Press.

Attfield, R. (1991). *The Ethics of Environmental Concern*. London: University of Georgia Press.

Benhabib, S. (1995). The debate over women and moral theory revisited. In Meehan, J. (ed.). *Gendering the Subject of Discourse: Feminists Read Habermas*. London: Routledge.

Bhaskar, R. and Danermark, B. (2006). Metatheory, interdisciplinarity and disability research: a critical realist perspective. *Scandinavian Journal of Disability Research*, 8(4), 278–297.

Bhaskar, R. (1986). *Scientific Realism and Human Emancipation*. London: Verso.

Bhaskar, R. (1993). *Dialectic: The Pulse of Freedom*. London: Verso.

Bird-David, N. (1993). Tribal metaphorization of human-nature relatedness: a comparative analysis. In Milton, K. (ed.) *Environmentalism: The View from Anthropology*. London: Routledge.

CARE (2009). *Care in Guatemala: Reversing Land Degradation & Building Carbon Stocks*. Project report at www.careclimatechange.org.

Cohen, G.A. and Arato, A. (1994). *Civil Society and Political Theory*. Cambridge, MA: Massachusetts Institute of Technology.

Cuomo, C. (1998). *Feminism and Ecological Communities*. London: Routledge.

Curtin, D. (2005). *Environmental Ethics for a Postcolonial World*. Oxford: Rowman and Littlefield.

Della Porta, D. (2005). Multiple belongings, tolerant identities, and the construction of 'another politics'. In Della Porta, D. and Tarrow, S. (eds.). *Transnational Protest and Global Activism*. Lanham, MD: Rowman & Littlefield, pp. 175–202.

Drainville, A.C. (2004). *Constesting Globalisation: Space and Place in the World Economy*. London: Routledge.

Dwyer, S. (1998). Learning from experience: moral phenomenology and politics. In Bar On, B.-A. and Ferguson, A. (eds.) *Daring to be Good: Essays in Feminist Ethico-politics*, London: Routledge.

Eder, K. (1993). *The New Politics of Class: Social Movements and Cultural Dynamics in Advanced Societies*. London: Sage.

Fisher, B. and Tronto, J. (1991). Toward a feminist theory of care. In Able, E., Nelson, M. (eds.). *Circles of Care: Work and Identity in Women's Lives*. Albany, NY: SUNY Press.

Frazer, E. and Lacey, N. (1993). *The Politics of Community: A Feminist Critique of the Liberal-Communitarian Debate*. London: Harvester Wheatsheaf.

Gronemeyer, M. (1993). Helping. In Sachs, W. (ed.) *The Development Dictionary*. London: Zed Books.

Held, V. (1993). *Feminist Morality*, London: University of Chicago Press.

Held, V. (1995). The meshing of care and justice. *Hypatia*, **10**(2), 128–132.

Held, V. (2006). *The Ethics of Care: Personal, Political and Global*. New York: Oxford University Press.

Hines, C. (2003). Time to replace globalisation with localisation. *Global Environmental Politics*, **3**(3), 1–7.

Hopkins, R. (2001). *The Transition Handbook: From Oil Dependency to Local Resilience*. Totnes, UK: Transition Books.

Jasper, J.M. (1997). *The Art of Moral Protest*. London: University of Chicago Press.

Johnston, P. (1989). *Wittgenstein and Moral Philosophy*. London: Routledge.

Jonas, H. (1984). *The Imperative of Responsibility*. Chicago: University of Chicago Press.

Klein, N. (2007). *The Shock Doctrine*. London: Penguin.

Liu, J. (2009). Loess leader (environmental restoration) at http://www.forumforthefuture.org/greenfutures/articles/602724.

Lovibond, S. (1983). *Realism and Imagination in Ethics*. Minneapolis: University of Minnesota Press.

MacIntyre, A. (1984). *After Virtue*. Notre Dame, IN: University of Notre Dame Press.

Melucci, A. (1996). *Challenging Codes: Collective Action in the Information Age*. Cambridge: Cambridge University Press.

Meyer, A. (2000). *Contraction and Convergence: The Global Solution to Climate Change*. Dartington UK: Green Books.

Midgley, M. (1984). *Wickedness: A Philosophical Essay*. London: Ark.

Mies, M. and Shiva, V. (1993). *Ecofeminism*. London: Zed Books.

Niezen, R. (2003). *The Origins of Indigenism: Human Rights and the Politics of Identity*. London: University of California Press.

Outhwaite, W. (1999). The varieties of critique: the Frankfurt School, Paper at Critique and Deconstruction conference University of Sussex, July 1999.

Parker, J. (1998). The precautionary principle. In Chadwick, R. (ed.) *Concise Encylopaedia of the Ethics of New Technologies*. London and San Diego: Academic Press.

Parker, J. (2001). Social movements and science: the question of plural knowledge systems. In Lopez, J. and Potter, G. (eds.) *After Postmodernism: An Introduction to Critical Realism*. London: Athlone Press.

Parker, J. (2004). How might the inclusion of discursive approaches enrich critical realist analysis? The case of environmentalisms. In Joseph, J. and Roberts, M. R. (eds.) *Realism, Discourse and Deconstruction*. London: Routledge.

Plant, J. (1990). Searching for common ground: ecofeminism and bioregionalism. In Diamond, I. and Orenstein, G.F. (eds.) *Reweaving the World*. San Francisco: Sierra Club Books.

Plumwood, V. (1993). *Feminism and the Mastery of Nature*. London: Routledge.

Poole, R. (1991). *Morality and Modernity*. London: Routledge.

Rapp, R. and Ginsburg, F. (2004). Enabling disability: rewriting kinship, re-imagining citizenship. In Scott, J. W. and Keates, D. (eds.) *Going Public: Feminism and the Shifting Boundaries of the Private Sphere*. Champaign: University of Illinois Press.

Sachs, W. (1999). *Planet Dialectics*. London: Zed Books.

Sandel, M. (1982). *Liberalism and the Limits of Justice*. Cambridge, UK: Cambridge University Press.

Scott-Cato, M. (2008). *Green Economics*. London: Earthscan.

Seddon, J. (2008). *Systems Thinking in the Public Sector: The Failure of the Reform Regime . . . And a Manifesto for a Better Way*. Axminster, UK: Triarchy Press.

Shiva, V. (2008). *Soil not Oil*. Boston, MA: Southend Press.

Society for Ecological Restoration (2007). *Ecological Restoration: A Global Strategy for Mitigating Climate Change*. Press release at www.ser.org/content/ecological restoration.

Solomon, R.C. (1980). Emotions and choice. In Rorty, A. (ed.) *Explaining Emotions*. Berkeley: University of California Press.

Soper, K. (1995). *What is Nature?* Oxford, UK: Blackwell.

Stammers, N. (2008). *Human Rights and Social Movements*. London: Pluto Press.

Stammers, N. and Eschle, C. (2005). Social movements and global activism. In De Jong, W., Shaw, M. and Stammers, N. (eds.) *Global Activism. Global Media*. London: Pluto Press, pp. 50–67.

Starhawk (1988). *The Spiral Dance*. San Francisco: Harper and Row.

Taylor, C. (1995). *Philosophical Arguments*. London: Harvard University Press.

Touraine, A. (1981). *The Voice and the Eye: An Analysis of Social Movements*. New York: Random House.

Tronto, J. (1995). Care as a basis for radical political judgements. In *Hypatia*, **10**(2), 141–149.

Walby, S. (1999). Against epistemological chasms: the science question in feminism revisited. Paper at the conference of the International Association for Critical Realism, Orebro.

Wendell, S. (1996). *The Rejected Body: Feminist Reflections on Disabilit*. London: Routledge.

Wollstonecraft, M. (1929). *A Vindication of the Rights of Woman*. London: Everyman.

Wood, E.M. (1995). *Democracy Against Capitalism: Renewing Historical Materialism*. Cambridge UK: Cambridge University Press.

13 Epilogue: the travelling circus of climate change

A conference tourist and his confessions

Karl Georg Høyer

Jet-propelled academics

> The whole academic world seems to be on the move. Half the passengers on transatlantic flights these days are university teachers. Their luggage is heavier than average, weighed down with books and papers – and bulkier, because their wardrobes must embrace both formal wear and leisurewear, clothes for attending lectures in, and clothes for going to the beach in, or to the Museum, or the Schloss, or the Duomo, or the Folk Village. For that's the attraction of the conference circuit: it's a way of converting work into play, combining professionalism with tourism, and all at someone else's expense. Write a paper and see the world!

These are the words – published more than 20 years ago – by the English author David Lodge (Lodge, 1984, p. 231). In several books he has, with both insight and humour, written about conference tourism, mostly expressed through the travels and experiences of a notorious conference tourist, the American Professor Morris Zapp (Lodge, 1975, 1984). He is zapping, from conference to conference; a real *conference-zapper*, a jet-propelled academic ever on the move. Lodge writes:

> The modern conference resembles the pilgrimage of medieval Christendom in that it allows the participants to indulge themselves in all the pleasures and diversions of travel while appearing to be austerely bent on self-improvement. To be sure, there are certain penitential exercises to be performed – the presentations of a paper, perhaps, and certainly listening to the papers of others. But with this excuse you journey to new and interesting places, meet new and interesting people, and form new and interesting relationships with them; exchange gossip and confidences (for your well-worn stories are fresh to them, and vice versa); eat, drink and make merry in their company every evening; and yet, at the end of it all, return home with an enhanced reputation for seriousness of mind. Today's conferees have an additional advantage over the pilgrims of old in that their expenses are usually paid, or at least subsidised, by the institution to which they belong, be it a government department, a commercial firm, or, most commonly perhaps, a university.
>
> (Lodge, 1984, Prologue)

We should listen, perhaps more, to fiction authors. We may have a lot to learn. In my field of research, the difference between research literature and fiction literature is not that large. As authors, we are often very close; building on own experiences and observations, utilizing and combining various types of sources, and wishing to contribute to a better understanding of the world around us, which we all are part of. We, the research academics, might call it interpretation. They call it something else. As regards methodical work, clever fiction authors may be just as good as clever research authors. We should beware that a critical view on the late modern lives of academics, not the least their jet-propelled conference tourism, has become an important topic also of many in the new generation of authors. And they, as David Lodge, are very often expressing their own experiences. A young Norwegian author, Helene Uri, has for instance lately written a humorous novel – *The Best Among Us* – showing outstanding insight into this world, a world she herself once was part of (Uri, 2006, p. 61):

> Most academics think giving a talk at a conference is quite exciting even though they are used to public speaking, even though it is painfully obvious to them that the audience has precious little interest in the content and even though most talks have been given before, at other conferences, in other towns and in other countries. When he started, just after he got his bursary, [he] had been shocked to find out that university staff travelled the world to conferences, holding the same talk time after time, to each other. The same witticisms, the same Power Point presentation or the same 'handout' (the Nordic purists insisted on their own Nordic term for this). The titles of the talks would change of course, so that one could apply for and reasonably expect to receive funding and have the talk registered on the research database at the mother institution upon one's return. As the years have passed, the shock has turned into resignation; now his reaction is to observe and swear an oath that he will always continue to rewrite his talks, or at least radically revise them, in order that he can consider them as original work with barely a pang of conscience.

And I myself have been on a conference tour again. To Shanghai. Of all things, to present a paper on the developments and problems of conference tourism, a special form of tourism in late modernity. It was a World Congress, of something. This summer, in July, I had actually been invited to two world congresses. Rightly, the other one was termed a *Symposium*: a World Symposium on Alcohol Fuels. Quite something, if we remember that the old Greek meaning of symposium is a boozing session combined with telling witty stories. And the programme folder had indeed been very tempting, if not exactly witty. The whole front page was a picture of the view from the bath tub in one of the conference hotel rooms, looking towards the beautiful beaches on Phuket Island in Thailand, where the symposium was going to take place. All hotel opportunities and pre- and post-conference tours were described on the second page, while one had to get to the third page to find a very short description of the actual conference programme. Of course, *sustainability* was an important topic in both congresses; in Shanghai it was sustainability both in planning and tourism, while it was sustainable energy and transport in Thailand. Still, my choice was Shanghai. After all, it seemed to be more stimulating for a university researcher. I had never been to mainland China before.

Excursus 1: Two different members of the conference tourist profession

I met him in Hong Kong. At another International Conference. With Sustainable Transportation as the head topic. I had seen him at the dinner for honourable guests the night before. He was from one of the Nordic countries. We had lunch together, after the conference opening with keynote addresses and a plenary session, where I had one of the presentations. The various sessions were going to start after lunch. But he wasn't going to participate in any of them. He had been to a conference in Hong Kong before, and he knew about a shopping area where he could buy high quality but cheap shoes. And this was where he was going to spend the rest of the day. And he was indeed very frank about his further plans. Last time, he did not have the opportunity to go to nearby Macao. He was not going to miss it this time.To go there was his plan for the whole next day. He spoke vividly about Macao, mostly known for all its casinos. Neither did he turn up the third and last day of the conference. Not even for lunch. But then, of course, he wasn't alone. As always in international conferences very few attend the closing sessions and summing up plenary. For tourist conferees, there is always some shopping, or sightseeing, to carry out in the last hours before the planes leave for home.

From a former participant in *the travelling circus of climate change* I can quote (Tveitdal, 2008, p. 139):

> The international environmental politics is big in one area: the number of meetings and air travels. My clever boss at UNEP in Nairobi, Klaus Töpfer, formerly Minister of Environment in Germany, was without competition in his efforts on the global environmental arena. In his persistent eagerness to participate in all these meetings he became number one air traveller in the loyalty program for air passengers within the Dutch air company KLM. However, we could not be satisfied with our results when the ambition was to make a turn towards sustainable development in global environmental development.

Call-girls

Lodge published his first novel about Morris Zapp and conference tourism, *Changing Places*, in 1975 (Lodge, 1975). But only a few years earlier Arthur Koestler (1972) published a book devoted to the same subject. It was titled *Call-girls*, also the term he gave the notorious conference tourists. And, as with Lodge, Koestler really has something important to tell us, about ourselves and how we are considered by others. The call-girls are the conference tourists. They are a regular group of individuals, an exclusive network of 'girls', predominantly men, indeed, who are called for one event after another. In his book, Koestler describes an international symposium day by day, but also each single call-girl's trip to and from the picturesque Alpine village where the event takes place. It is a description that hits just as much today.

Koestler writes (text concentrated by this author):

> One of the most accepted rituals for all congresses, conferences, symposiums and seminars is the get-to-know cocktail party the evening before the start of the formal program. Indeed, getting to know was hardly necessary, as most of the participants already knew each other from similar events elsewhere. Most of them, with a few exceptions, arrived precisely for the cocktail. The participants included wives as well as staff members of the secretariat. Memories and points of view from the last time when they met were exchanged. Most of them seemed totally uninterested in the magnificent view to the Alp landscape. As usual, the whole thing started a bit formally, but everyone knew that it would gradually become quite noisy and unrestrained, for some also followed by a lovers' hour in the hotel room.
>
> This was on Sunday. At 9 o'clock sharp on Monday morning the participants were situated along the conference table, each with a notepad and a file. The latter was supposed to include abstracts of all papers, but as usual, most participants had not been able to complete any paper or abstract in time. Indeed, some of those expected to come did not show up at all. This always happened with call-girls. Some of them were always late, some had to leave before it was over, some came only for a day, delivered their presentation, had their expenses covered and got their fee, and then travelled on to the next event. On Tuesday, one of the participants, in a moment of critical reflection, could see the connection between the stupid-making mass tourism and the call-girls – between the tourist explosion and the knowledge explosion – and the polluting fallout left by both in the societies they visit so briefly. On the third day of the symposium – Wednesday – the expected duel between two of the participants had been performed as expected by everyone. In fact, it was not their first confrontation, they had already met and quarreled twice before that year – at an ecology congress in Mexico City and at a futurology symposium at the Academy of Stockholm.

Excursus 2: Movement and change

There are necessary relations between movement and change. Changes in means, patterns and levels of human movement are inter-connected with changes in tourism. These changes have ecological effects. They are in a particular part of the global processes causing climate change. A basic understanding of mine is that there is nothing like a neutral movement. And there is no physical or ecological neutrality. Illusions of such movement neutralities have a long history. In almost any traditional culture in the world there are old fairy tales about how long distances can easily be overcome through 'flying carpets' or '7-mile boots'. No limits, no change, and no impacts. Such adventurous tales are also well known from my own country – Norway. It is, of course, quite another thing to try to make the adventures become reality, actually quite an appropriate description of the current global development in movements and tourism. There are limits to the extent of movements. And there are the inter-related limits to the extent of tourism (Høyer and Aall, 2005).

This separation between movement and change has a place in the very foundation of modernization. The scholars of the age of Antiquity did not draw the line in this way. Notable were the physics and biology of Aristotle. To him, movement and change were one thing; all forms of changes were understood as forms of movement. Not only did this apply to movements as such, but also to growth and even to changes in colour in nature. The physics of Galileo and Newton was in contravention to this view. Their programme could not be fulfilled if all changes were considered to be based on one common principle. They had to separate out the physical movements between places and points. This should become an important part of the new modern worldview, and the prevailing process of modernization which we still are part of (Hägerstrand, 1993; Høyer and Aall, 2005).

A crucial concept of mine is mobility and not movement. It is a concept very much highlighted in recent sociology (Urry, 2000, 2003, 2007; Kaufman, 2002; Scheller and Urry, 2004). Without mobility of humans there is no *tourism*. It lies in the very etymological origin of the word 'tour'. The origin is the Latin *tornare* and the Greek *tornos*, which is the movement in a circle around a central point or axis. The word tourism, then, actually means taking a round trip. The core lies in the movement itself, away from the starting-point and back again (Høyer, 2000). This is clearly expressed in my country. In Norway, the term applied is *travel-life industry*. Tourism is a more limited category within this term. In the Norwegian language a tourist was, first of all, a rich foreigner. Originally you found them in upper-class hotels along the fjords of Western Norway, or driving in open seven-seat cars and in first-class train or coastal ship compartments. Today, however, *tourist class* in travelling is actually low class, and the use of the term is closer to Henry James' 'Tourists are vulgar, vulgar, vulgar'. This is a rather contradictory view as most of us have become tourists during this century of change (Høyer, 2000). As movement, mobility is also very much handled as a category in itself, separated from change. But in other contexts, and not the least in our daily language, we are closer to the understanding of the age of Antiquity. We apply terms like occupational mobility, population mobility and family mobility; all implying change, but often with physical movement as an important precondition.

Leisure time mobility

Relations between transport and leisure time have always been very close (Høyer and Aall, 2005). Even if today we find *leisure time mobility* very dominating, the phenomenon is not at all new. Leisure time activities, sport and tourism, have, so to speak, paved the way for the growth in various transport means. Horse transport only had minor importance in this relation. The 'new times' were first of all augured by the *bicycle*. A really improved type, with iron frame and pedals on the front wheel, was launched at the World Exhibition in Paris in 1867. It was called *velocipede*, with connotations to *velocity*, and soon became very popular, particularly after Dunlop introduced his rubber tyre in 1888. Sports and other leisure time uses were crucial in marketing the new transport, and bicycle sport arrangements had become very extensive by the late 1800s. All users were called *cycle riders*, to remind us of *horse riders*. And, of course, horse riding was mostly for sport and leisure time; the organized transport function of horses at that time was in the form of coaches. Today we only call them cyclists, but cycle sport and ever more indigenous leisure time use of cycles are just as important. Not least we are today talking of cycle tourism as a separate form of tourism. Some basically connect this to the concept of *sustainable tourism*.

But it has changed in very important ways: today, we transport the bicycles by car or we travel by planes to *enjoy*, as we say, the pleasures of cycle tourism. And many travel by car to health studios in order to bicycle inside, without going anywhere, a form of *virtual mobility* in late modernity. The velocipedes were pure marvels of speed, and all the cycling required higher quality roads than horse transport. Major road improvements were made both in urban and rural areas. The bicycle so to speak paved the way for the new means of transport, the *automobile* (Høyer and Aall, 2005).

Stronger efforts were, however, required to extend the use of cars. Most people did not need cars in their daily lives: they walked or bicycled to work and to nearby shops. New urban rail systems gave the opportunities for longer journeys. Thus, cars were neither needed for *production* nor *reproduction* related mobilities. Close links were, on the other hand, made between the car and a third category of mobility: *leisure time mobility*. Car-use, so to speak, started as a purely leisure time activity, and this link has become fairly prominent during the whole car-age history. Early advertisements presented cars as a means to go out into the fresh country air and landscapes, and away from the industrialized and polluted cities. This was even marketed as a health measure; while driving in open cars, one could breathe in fresh air and thus help to cure the tuberculosis caused by city life. *Sport* has, through all times, had an important marketing function. Car-*races* – with connotations to horse-races – were very soon set up (Høyer and Aall, 2005).

A life with aeromobility

In late modernity we belong to societies of *aeromobility*, just as they are societies of automobility (Høyer and Næss, 2001). This is a global mobility, which in extent and type has the aeroplane as a fundamental precondition, and at the same time conveys the historical illusion of movements without limits or impacts as in the fairy tales with flying carpets. It plays – as automobility – a major role in structuring late-modern societies, where leisure time and tourism are particularly important components. As a conference tourist I live a life in and with airplanes. It is a highly mobile life. It is worth remembering that *mobility* also implies something that is fluctuating, as in *Donna est mobile*, the words uttered by the count in the opera *Rigoletto*. And in the lives as conference tourists, the Donnas are mostly just as present, and mobile, as the Dons.

Aeromobility then is both: mobility without limits and with large fluctuations, but more extreme than in the case of Donna. The plane to Shanghai had an average speed, in the air, of about 1,000 km/h. But the speed of the luggage conveyor belts was less than 10 km/h. It is impressively easy to move some 9,000 km from Frankfurt to Shanghai on one flight. Everything that takes place on the ground is, on the other hand, utterly cumbersome. The ease to move in the air has necessary repercussions in the form of extreme compactness on the ground. This is where the old fairy tales about limitless mobility really were wrong. In Frankfurt airport I was told that the luggage would turn up on the conveyor belt at any time between 10 minutes and 2 hours. It did become closer to the 2 hours. Coming from Oslo, I changed plane in Copenhagen. I had to cross a large part of the airport area to hurry from one main terminal to another. And when I landed in Frankfurt, I also had to walk quite some distance. Actually these days, I never walk as much – and as hastily – as I do in airports. This is probably the case for very many of us, that we are forced to walk more than people used to do in former periods of industrial modernization. For the most part, these walks are caused by the traffic-related agglomerations, whether they are to and from parking lots for cars, inside shopping malls, along shopping streets for pedestrians, or from one air terminal to another. In the life as a conference-tourist, it is mostly walking within corridors. This thesis – *a life in corridors* – I shall return to.

Excursus 3: The mobilities of conference tourists

Figure 13.1 shows the daily mobilities of a typical conference tourist. He is a Norwegian, and his name is Karl. A comparison is made with an average Norwegian.

Both the levels and patterns of mobility are very much at odds with each other. Karl has a daily mobility which is two to three times more than the average Norwegian. This is caused by the extent of international as well as domestic air travel. Almost all of these journeys are connected to research conferences. His international conference mobility alone is much higher than the average Norwegian mobility for all purposes. Karl has a fairly low car-based mobility, and a higher than average public transport use. However, as part of the total figure this has only minor importance. And Karl has relatively high energy consumption for his car as he mostly drives alone, and mostly as leisure time driving.

Figure 13.2 compares the international aeromobilites for an average Norwegian, the conference tourist Karl, and a typical professional member of the travelling circus of climate change. The contrasts between the three are indeed very prominent, but particularly so with regard to the differences between the average Norwegian and the professional climate change traveller.

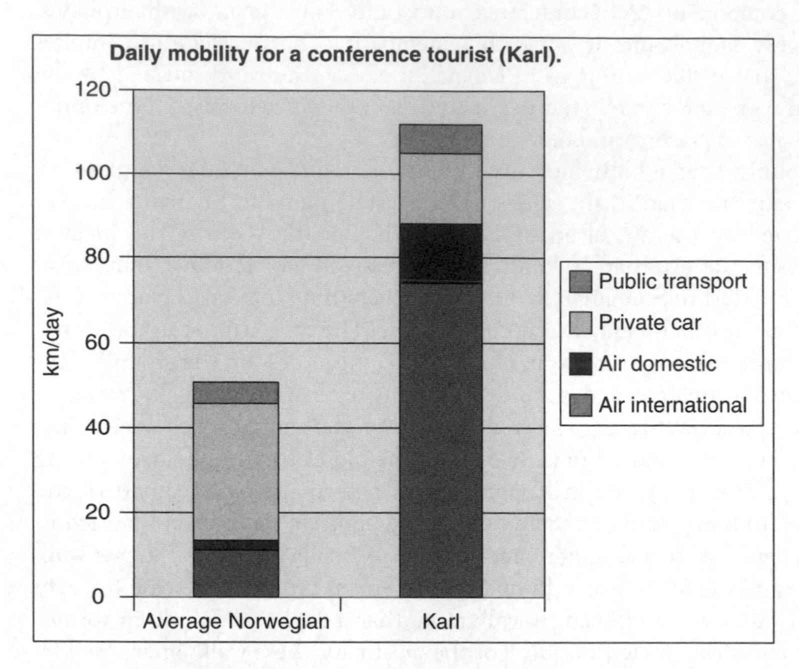

Figure 13.1 Daily mobility for a conference tourist (Karl). Comparison with an average Norwegian. 2006 – figures in km/day (Høyer, 2003; Hille *et al.*, 2008).

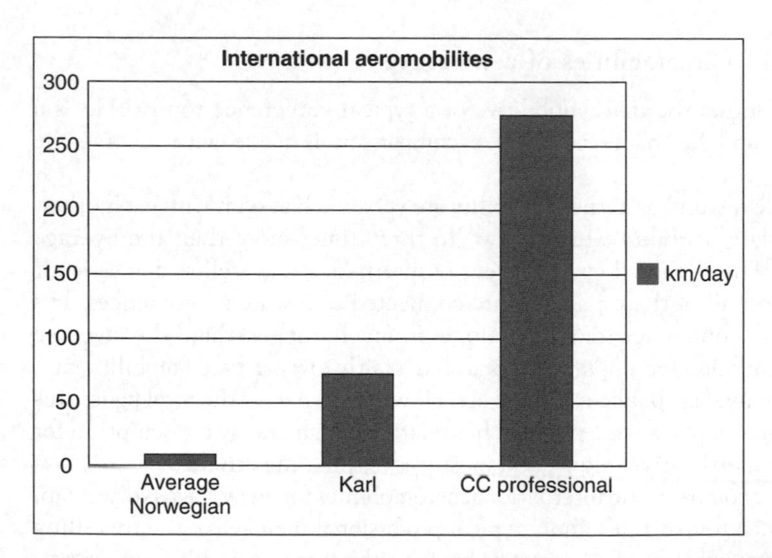

Figure 13.2 International aeromobility for a professional member of the travelling circus of climate change compared with an average Norwegian and the conference tourist Karl. 2006 – figures in km/day.

A life with compost-modernism

As a conference tourist I live a life with *compost-modernism*. I am concerned about environmental problems, locally and globally. I bike or walk to my job, have a well-insulated dwelling, recycle all waste, and even make a little *compost* in the garden. But I also pass through the local airport more than 20 times a year, mostly in order to participate in conferences on 'sustainable development' or give lectures on 'transport, tourism and the environment'. And the energy use and ecological effects of all the air travel are substantially larger and much more serious than all the local problems I try to solve with my presentations. A life with compost-modernism is a life between two extremes. On the one hand, the compost relation gives meaning as a very firmly founded way of life, anchored to the ground so to speak, and more like earlier forms of *fixed modernity*. On the other, all the travelling, far away through the air and only with temporary, superficial relations to the places visited, represents a most radical version of later forms of *liquid modernity*, as the sociologist Zygmunt Bauman (2000) termed it.

Excursus 4: Between liquid and fixed energy consumption

Figure 13.3 illustrates the daily energy consumption of a typical conference tourist. It is the Norwegian Karl. His consumption – and, of course, more so for a

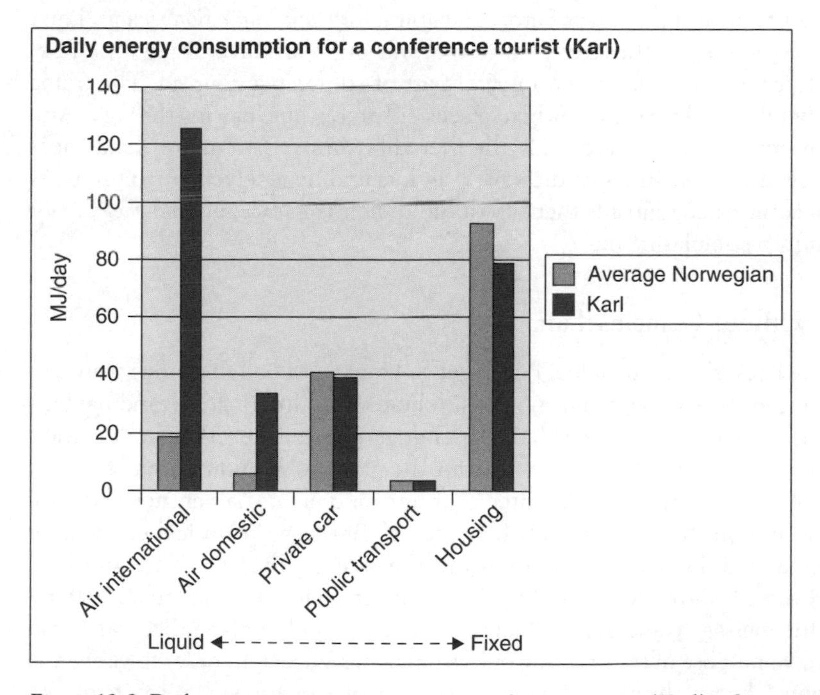

Figure 13.3 Daily energy consumption for a conference tourist (Karl). Comparison with an average Norwegian. 2006 – figures in MJ/day (Høyer, 2003; Hille *et al.*, 2008).

professional member of the travelling circus of climate change – is starkly in contrast with the average Norwegian. He has a much higher consumption towards the most liquid parts, international and domestic air transport. However, he still keeps the more fixed energy consumption, for housing and public transport, at the same level as the average. This gives an illustration of the term compost-modernism. The average Norwegian, on the other hand, has the highest levels of consumption towards the most fixed parts of the liquid-fixed continuum.

A life with traffic infarction

The extreme ease and speed of movements in the air causes bottlenecks to turn up all the time. They are everywhere; on the airport runways, in the air space above the airport, on the luggage conveyor belts, and on the highways to and from the cities. Many times the bottlenecks develop into incidences of *traffic infarction*. This is when the luggage gets lost and is returned only after I have got back home again. Or not least when delays in one place mean that I miss the connecting flight, and perhaps must stay overnight to wait for the next one. As a conference tourist, I live a life as a victim of traffic infarction. But then there are institutions to take care of me, just like hospitals for infarct patients. They are *traffic infarction hotels* that are specializing in giving food and shelter one night only for victims such as myself. Even Frankfurt city, as a major node in European air traffic, is characterized by its function as European capital for *traffic infarction victims*. I met them everywhere, on the streets, in restaurants and bars, all waiting for a next plane. It is a whole industry, an integral part of conference tourism. The traffic infarct hotels are like *motels*, nowhere places only for people passing through. And they are situated in nowhere lands, the incredibly unpleasant airport landscapes with their traffic machines, endless parking lots and huge advertising posters. A life as a conference tourist is then also a life in *nowhere places* and *nowhere lands*. It is hardly a stimulating life.

A life without Gemeinschaft

My plane left Oslo as scheduled. But to get to Frankfurt I had to change planes in Copenhagen. It is also a major node, like Frankfurt airport, for Scandinavians only. This is where they often have to change planes when they are on their conference tours. In order to be in time for the plane to Frankfurt, I had to hurry from terminal B to terminal A. After boarding, and about half an hour late, we were waiting on the runway ready for take off. But some technical errors were discovered, and the plane had to return to the gate. By then I was an hour late, and of course I started to worry about my connecting flight to Shanghai. After a while, the message came that we had to change to another plane that was set up for the sole purpose of transporting our group of passengers. In order to get to the new gate, I had to hurry, walking across most of Copenhagen airport for the second time. When all passengers had boarded, I was already delayed to the extent that I would miss my Shanghai plane. New delays would, however, occur. With all passengers seated, the message from the pilot was that we now were in the hands of the air traffic control officers at Frankfurt airport. Due to traffic queues in the air space above Frankfurt we were told that we would have to stay another 2 hours in the plane before take-off.

But then suddenly there was a contra message from the captain. The queue in the air had been resolved and we were allowed a fast take-off from Copenhagen, about 3 hours late. Ironically, it was exactly the same time as the plane to Shanghai was to leave Frankfurt. And, even more ironically, the sign on the first van I saw on the airport after landing was: 'Airport Service Gesellschaft – Just-in-time'. In themselves, terms belonging to a world of liquid modernity; strong as symbols but weak as realities, and laden with words from the particular Anglo-American air transport language. The sign reminded me of the distinction made in 1887 between *Gesellschaft* and *Gemeinschaft* by the German sociologist Ferdinand Tönnies. In coherent societies, marked by Gemeinschaft, people are intrinsically linked to each other both through their dispositions and ways of behaving, while in societies marked by Gesellschaft, people associate with each other in rational and calculating ways. From the outside they may look just as coherent, but from the inside they are divided. Airports – and airplanes – no doubt belong to the Gesellschaft type of society, just as the sign on the van told me.

Then, as a member of the Gesellschaft society, I managed to walk faster than most of my co-passengers towards the nearest Lufthansa desk, and queued up as one of the first. I received a boarding card for the Shanghai plane the next day and a hotel coupon to a nearby hotel. From the luggage information, I was told that the luggage would turn up on the conveyor belt anytime between 10 minutes and 2 hours. It did become closer to the 2 hours. The queuing of luggage on the ground was obviously just as normal as the queuing of planes in the air.

A life as a left-over

The hotel was a ten-storey building not far from the airport, situated along one of the highways connecting the city to the airport. Together with a lot of other hotels in the area it was an integral part both of the airport structure and its surrounding landscape, the international airport landscape. It was an incredibly unpleasant landscape, a *nowhere land* full of monotony and huge advertising posters with their standardized symbols and airport language. The hotel itself was not any more pleasant. There was no doubt whatsoever what type of guest I was. They did not ask whether I had booked, and only wanted to see the coupon from the airline company. I received three new coupons, one for dinner and the other two for breakfast and lunch. Everywhere there was a queue, and all the meals were standardized with a limited number of dishes consisting of a few standard components. It was a hotel solely specializing in the left behinds from air transport delays: the victims of *air traffic infarction*. The hotel did not do anything to make guests wish that they could come back; they could always expect new groups of left behinds the next day. Similarly, none of the guests would ever want to come back, unless forced to. It was like a *motel*, not linked to automobility however, but to aeromobility. And like other motels, it was a *nowhere place*, a place only to pass through.

As a term in social sciences *critical theory* originated in Frankfurt with the Frankfurt school. The school, a unique circle of academic scholars, was most productive during the 1930s and 1940s. But it was also highly active during the 1950s and 1960s, and had its most profound political effect when giving inspiration to the student revolts in France in 1968. Of course, these revolts later spread to most Western countries, and inspired large groups of leading intellectuals. Jürgen Habermas, still living in Frankfurt, was a member of the school during its later period. His philosophy and social theories have a similarly strong influence today. A critical question raised by the Frankfurt school was why the seemingly positive development of technical productive forces led to such an extensive development of destructive forces, and why this could take place both in late capitalistic and socialist countries. How was it that humanity in its height of technological development opted for roads that implied barbarism? They claimed that a crucial point was the relationship between humans and nature in all Western European types of civilization, and that the main reason was to be found in the conquering and exploiting mastery over nature so dominating in this civilization.

A life in complication

There is much talk about complexity these days (Urry, 2003). Terms like social complexity, urban complexity, global complexity are applied in post-modern social theory. Their application of the basic complexity term is up for discussion (Høyer, 2007). Within Norwegian ecophilosophy, the distinction between complexity and complication was made already in the early 1970s, particularly through the works by the Norwegian philosopher Sigmund Kvaløy (1973). Complexity is organic, diverse coherence as it is in ecological complexity. As in system ecology it is *diversity in structure, function and communication*. Complication, on the other hand, is without this organic coherence, even though a form of diversity may be present. A distinction can be made between artificial and organic diversity, a distinction that is not readily grasped in post-modernistic analyses. Complication is thus an expression for the multi-factored set of entities and relationships that can be analysed in quantitative terms. Complexity involves operating on *many qualitative levels, besides* the level of complication.

Complication was introduced in the world by the human mind, and refers to the special cluster of difficulties and, finally, impossibilities inherent in the attempt to grasp and control complexity through quantification. It is this one level of theorizing and organizing which has given man his industrial and technological might, and in the Industrial Growth Society (IGS) it is replacing all the other levels of relating to our total environment. And the 'Apparatlandschaft', where the society dominantly masters its total environment through technologies, contributes to the dominance of the level of complication (Høyer, 2007).

Kvaløy (1973) compares human life in the 'Apparatlandschaft' with life in a *pipe system*, another of his crucial concepts. The most functional citizens of the IGS society are those with no protruding edges, those who have the slippery shape of simple discs, well suited to slide through the centrally designed pipe system of the modern technosociety.

If, however, this pipe system should break down, for instance, through drastic ecological changes, then the standardized pipe structure and the lack of individual and social complexity would show up as extreme vulnerability. The majority, trained to function as simple discs, would then reveal their complete dependence on the tubular guidance of the pipe system to which they have been fitted. All of the energy of the 'disc-humans' is used up for one purpose: to manage and communicate *complication*. They move easily within the pipe system, but are lost outside or without it.

A life in corridors

Inspired by the ecophilosophy term 'living in pipes', I have used the concept or metaphor of *life in corridors* in my analyses of the lives of late modern, global conference tourists in academia (Høyer, 2007). A corridor is a long narrow aisle, only with a lot of closed outlets, doors, to the sides. Corridor politics is political negotiations being carried out within closed circles. It has negative connotations, being the very opposite of open, democratic procedures. During the past few decades it has been much used in transport planning and politics. We talk about *transport corridors*. As in the historical term the *Polish corridor*, they are channels established through foreign territories to achieve highly efficient communication, but there is only communication along the corridor; it is closed to both sides. There is no communication to the territories they pass through. This is the very condition for efficiency. And, of course, they have the asset of being collective routes for very many. However, they are not only without communication to the sides; everyone following the corridor is doomed to come to the same corridor end, the ultimate destinations, as was the case with the Polish corridor.

In the life of a conference tourist there are corridors everywhere. There are corridors along the highways, or high speed railways, to the international airports for departure. For the luggage on transport belts, following all the others to the departure gates. Yes, this is exactly what they are called: *gates*. There is no airport term as important as this. Then, going through the gates for embarking. Follow the others to your seat along the very narrow plane corridor. The airplane itself follows an air corridor, be it in the troposphere or the stratosphere. When landing, everything goes in reverse. This is seemingly an incessant forwards and backwards, just as we have learnt in Newtonian physics.

And you end up in an international hotel after having been carried along new corridors from the international airport you arrived at. But everything looks the same wherever you are in the world. It is certainly disenchanted. The airports are alike. The airport landscapes with their connected transport corridors look the same. The planes are the same. The international hotels are all alike. Even the language is the same: the globalized *Anglo-American*, the *corridor language* for the whole world. A standardized language of nothingness. This same language is also the *conference language* itself, where researchers gather to present their latest findings within the language turn in discourse analysis.

Enchanting a disenchanted world, cathedrals of consumption and climate change crisis

These are some concepts launched by the American sociologist George Ritzer (1998, 1999, 2004b). Consumption in late modernity takes increasingly the form of hyperconsumption. To serve this form of consumption, new means of consumption are developed, and globalized, and come to play important roles in *grobalization*. They are becoming much more dominant than the means of production, in the Marxian meaning. According to the Ritzer analysis the new means of consumption should be seen as cathedrals of consumption. They have an enchanted, sometimes even sacred, religious character for many people. And the journey to them seems just like that of a modern-day pilgrimage. It has been contended by several analysts that shopping malls for instance are more than commercial and financial centres; they have much in common with the religious centres of traditional civilizations (Ritzer, 1999, p. 8).

The main cathedrals of consumption are: super malls, shopping centres, casinos, cruise ships, amusement parks, major railway stations, football and athletic stadiums, but also many universities, museums, restaurants (*eatertainment*), the internet, and to really close the circle, even new mega churches are going through a similar development. And, not least, the major international airports are cathedrals of consumption. In order to attract ever-larger numbers of consumers, such cathedrals of consumption need to offer, or at least appear to offer, increasingly magical, fantastic and enchanted settings in which to consume. A type of enchantment applied is to make the shopping buildings ever larger. Not only malls, but now mega malls. Not only airports, but mega airports. They are planned to stimulate and force you to hyperconsumption, and they are, of course, the very backbones in the aeromobile life as hyperconsumers. In these airports you are forced to go along very long corridors with only a lot of shops to the sides. They have just rebuilt the international Oslo airport so that all international travellers are forced to go through a large tax-free shopping mall.

Nordic ecophilosophy emphasizes the importance of the *Disneyland effect* as a means to compensating the human tragedies of the industrial growth society; this is their description of the process of continuously enchanting a disenchanted world. But Ritzer extends this: there is a close relation between hyperconsumption and enchantment. The new cathedrals of consumption play crucial roles both as places for enchantment and as arenas for hyperconsumption. We do know very well from current development trends in our affluent societies that not only is shopping fun, it is one of the most fun things that we can do. But the effects on ecology are not funny. The cathedrals are closely knit to the particularly troublesome systems of hypermobility; the auto- and aeromobility. And all the commodities are produced from something, and take something – rather problematic – back to nature while being produced, consumed and when becoming waste.

A life with grobalization

George Ritzer (2004a,b) has launched the concept *grobalization* in contrast to the *glocalization* theses within globalization theory. Grobalization is the process in which growth imperatives, e.g. the need to increase sales and profits from one year to the next in order to keep stock prices high and growing, push organizations and nations to expand globally and to impose themselves on the local. He considers grobalization to consist of three major subprocesses: Capitalism, McDonaldization, and Americanization. It is a concept greatly at odds with the dominant postmodernistic theses of glocalization.

Ritzer thus underlines that there is a gulf between those who emphasize the increasing grobal influence of capitalistic, Americanized, and McDonaldized interests, and those who see the world growing increasingly pluralistic and indeterminate. Largely as antitheses to the glocalization theory, *grobalisation* is, according to Ritzer, seen to encompass the following: the world is growing increasingly similar. Individuals and groups have relatively little ability to adapt, innovate and manoeuvre within a grobalized world. Larger forces and structures tend to overwhelm the ability of individuals and groups to create themselves and their worlds. Grobalization tends to overpower the local and limits its ability to act and react, let alone act against the grobal. Commodities and the media are the key forces and areas of cultural change, and they are seen as largely determining the self and groups throughout the grobalized areas of the world.

In relation to this last point, *glocalization theory* (Robertson, 1994, 2001) also sees commodities and the media as key forces in cultural change in the late twentieth and early twenty-first centuries. However, in stark contrast to the grobalization position, it considers these forces as providing material to be used in individual and group creation throughout the glocalized areas of the world.

The grobalization theory of late modernity is very much in line with the analyses and descriptions carried out by the Norwegian ecophilosophers some 30 years ago, in particular as it has been elaborated in the works of Sigmund Kvaløy (1973). And grobalization forces expanding and imposing on the local generate severe ecological imbalances and environmental problems, locally as well as globally. This is not included in the Ritzer theory. To him, nature is largely non-existent; in this he follows a long tradition in sociology, carrying on the inheritance both from Durkheim and Weber (Høyer, 2007).

A life with globalization of nothing

Ritzer (2004a) also launches the thesis of globalization as the production of *nothing* and *nothingness*. *Grobalization* leads to an increasing dominance of nothing in the form of non-places, non-things, non-people, and non-service, all at the expense of something on a nothing–something continuum. All of us do, of course, 'meet' non-people every day, through internet communication, and when trying to talk to speaking machines on the phone. Non-places of late modernity are, for example, major highway crossings, highway motels and international airports. Here, to a large extent, we meet non-things, non-people, and non-service. The French anthropologist Marc Augé (1995) defines a *non-place* as a space which cannot be defined as relational, or historical, or concerned with identity. To describe such places he uses terms as solitary, fleeting, similitude, anonymity, lack of history, etc. Ritzer (2004a, p. 20) connects the following main characteristics to the nothing–something continuum:

Something	*Nothing*
– Unique	– Generic/Interchangeable
– Local geographic ties	– Lack of local ties
– Specific to the times	– Time-less
– Humanized	– Dehumanized
– Enchanted	– Disenchanted

According to Manuel Castells (1996) we are, in the so-called network society, moving from a world characterized by 'spaces of places' to one dominated by 'spaces of flows'. There is a high correlation between spaces of places and something, on the one hand, and spaces of flows and nothing on the other (Ritzer, 2004a, p. 41).

Norwegian ecophilosophy puts emphasis on the production of nothing, non-places and non-people. But within the late modern sociology, terms like spaces of flows, nothingness and non-places often convey the post-modernistic illusion that this is totally *foot-loose* in relation to nature; that these are processes that also represent nothing in relation to nature. This is quite the contrary. There is nothing that may imply ecological neutrality. Production of nothing definitely leads to something in nature, in even more severe forms than before. The global flows are fortifying processes of ecological crisis. Non-places are integrated in particularly environmentally harmful societal processes, as *hypertourism*, *hyperconsumption* and *hypermobility* in the form of extreme automobility and air-mobility, or aeromobility as I have termed it.

Excursus 5: The CO_2 emissions of conference tourists

The societal aspects of conference tourism may belong to the nothing end of the something–nothing continuum. But it leads to something; something that entails physically concrete ecological footprints. Figure 13.4 shows the daily emissions of CO_2 for all journeys made by the conference tourist Karl. Comparisons are made with the average Norwegian. The differences between the two are indeed very stark. More so than in the case of mobility and energy consumption, as they have been visualized in Excursus 3 and 4. The reason is that the ecological effect, the climate change effect, is relatively larger from emissions high up in the atmosphere than on the ground. This effect is called *climate forcing*, and has been

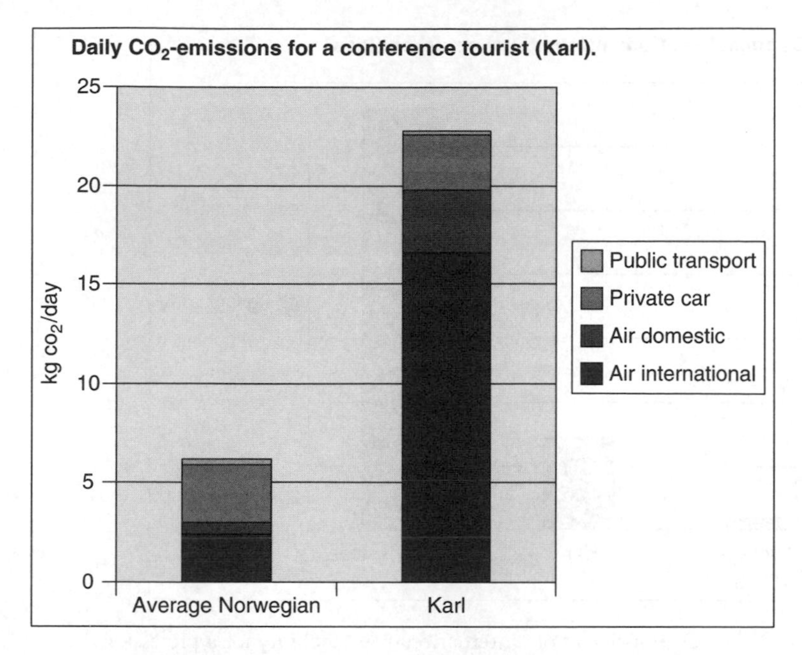

Figure 13.4 Daily emissions of CO_2 for a conference tourist (Karl). Comparison with an average Norwegian. 2006 – figures in kg CO_2/day includes climate forcing multipliers for air travels (Høyer, 2003; Hille *et al.*, 2008; Schlaupitz, 2008).

estimated through particular indexes or multipliers. The climate forcing multiplier is estimated to be 1.8 for international air travel, and 1.3 for domestic air travel in the Norwegian context (Hille *et al.*, 2008; Schlaupitz, 2008). Corresponding figures for emissions of CO_2 have thus been multiplied by these factors. The implication is a climate change forcing effect about *four times* higher for the conference tourist Karl than for an average Norwegian, while the difference in mobility is a factor of 2–3 as illustrated in the former Excursus.

Figure 13.5 compares the average daily emissions of CO_2 from international aeromobilities for an average Norwegian, the conference tourist Karl and a typical professional member of the travelling circus of climate change. The contrasts between the three are indeed very prominent, but particularly so as regards the differences between the average Norwegian and the professional climate change traveller.

An afterthought: a happy conscience?

About a year ago (2008), I was invited to another international conference. This time it was a European COST network meeting. As you might know, COST is an intergovernmental framework for researchers from all European countries,

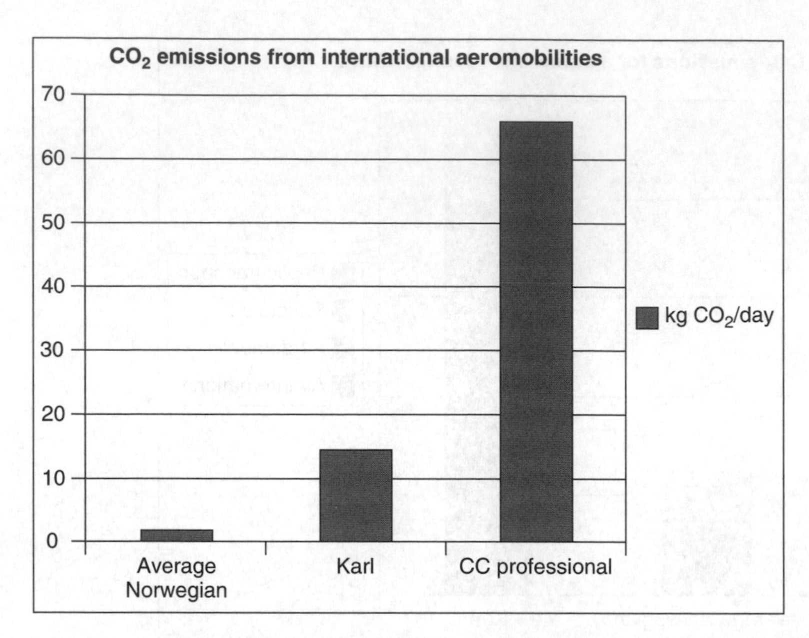

Figure 13.5 Daily CO_2 emissions from international aeromobility for a professional member of the travelling circus of climate change. Compared with an average Norwegian and the Conference Tourist (Karl). 2006 – figures in kg CO_2/day.

financed by the European Science Foundation. However, the meeting was to be held in Algeria, not in Europe. And not in the capital Algiers on the northern coast, but in Ghardaïa further south towards the Sahara desert, 600 km by plane each way from the capital. Ghardaïa is known to be a very exciting city, for tourists. I sent all participants an e-mail, where I explained why I didn't have the conscience to attend. I was the only one. Perhaps a bit astonishing as the common network research focus was '*sustainable mobility*'. In strict terms it was only *indicators* for sustainable mobility. But surely a focus on indicators doesn't relieve us from taking responsibility for the substantive implications of a concept.

Earlier this year, the message came that the Ghardaïa meeting had been cancelled. Was this due to a bad conscience after all? No, the land area had been subject to a severe natural catastrophe. Exceptional heavy rainfalls for many days, with flooding, and destructive impacts both on the inhabitants and their infrastructures. But 'natural', is that the right term? It may be just as fair to claim the cause as human made climate change. And the really profound tragedy is that the most important factor behind it may be CO_2 emissions, from global tourism and global conference tourism. And when the conference was cancelled, we all, as researchers, were relieved from the confrontation with this tragedy in real life.

References

Augé, M. (1995). *Non-places: Introduction to an Anthropology of Supermodernity*. London: Verso.

Bauman, Z. (2000). *Liquid Modernity*. London: Polity Press.

Beck, U. (1995a). *Ecological Enlightenment*. Atlantic Highlands, NJ: Humanities Press.

Beck, U. (1995b). *Ecological Politics in an Age of Risk*. Cambridge: Polity Press.

Castells, M. (1996). *The Information Age, 1. The Rise of the Network Society*. Oxford: Blackwell.

Hille, J., Aall, C. and Holden, E. (2008). [*Miljøbelastningen fra norsk forbruk og produksjon – 1987–2007*] *Environmental Impacts from Norwegian Consumption and Production – 1987–2007*. Report 2/2008. Sogndal, Norway: Western Norway Research Institute.

Hägerstrand, T. (1993). *Mobilitet*. Teknikdalen, Borlänge, Sweden: Johan Godtlieb Akademien.

Høyer, K.G. (2000). Sustainable tourism or sustainable mobility? The Norwegian case. *Journal of Sustainable Tourism*, **8**(2), 147–161.

Høyer, K.G. and Næss, P. (2001). Conference tourism: a problem for the environment, as well as for research? *Journal of Sustainable Tourism*, **9**(6), 451–471.

Høyer, K.G. (2003). [*Husholdninger, miljø og forbruk*] *Households, Environment and Consumption*. Documentary Report. Sogndal, Norway: Western Norway Research Institute.

Høyer, K.G. and Aall, C. (2005). Sustainable mobility and sustainable tourism. In C.M Hall and J. Highham (eds.) *Tourism, Recreation and Climate Change*. Aspects of Tourism series. Clevedon, UK: Channel View.

Høyer, K.G. (2007). Ecophilosophy and the Contemporary Environmental Debate. [*Sosiologisk Årbok*] *Yearbook of Sociology*, Norway, 3–4, 2007. Oslo: Novus.

Kaufmann, V. (2002). *Re-thinking Mobility. Contemporary Sociology*. Aldershot, UK: Ashgate.

Koestler, A. (1972). *The Call Girls*. London: Hutchinson.

Kvaløy, S. (1973). [*Økofilosofisk Fragment IV*] *Ecopilosophical Fragment IV*. Trondheim, Norway: Tapir.

Lodge, D. (1975). *Changing Places*. London: Penguin.

Lodge, D. (1984). *Small World*. London: Penguin.

Ritzer, G. (2004a). *The Globalization of Nothing*. London: Sage.

Ritzer, G. (2004b). *The McDonaldization of Society*. Revised New Century Edition. London: Pine Forge Press.

Ritzer, G. (1998). *The McDonaldization Thesis*. London: Sage.

Ritzer, G. (1999). *Enchanting a Disenchanted World: Revolutionizing the Means of Consumption*. London: Pine Forge Press.

Robertson, R. (1994). Globalisation or glocalisation? *Journal of International Communication*, **1**, 33–52.

Robertson, R. (2001). Globalization theory 2000+: major problematics. In G. Ritzer and B. Smart, B. (eds.) *Handbook of Social Theory*. London: Sage.

Schlaupitz, H. (2008). [*Energi- og klimakonsekvenser av moderne transportsystemer*] *Energy and Climate Impacts of Modern Transport Systems*. Report 3/2008. Oslo: Friends of the Earth Norway.

Sheller, M. and Urry, J. (eds.) (2004). *Tourism Mobilities. Places to Play, Places in Play*. London: Routledge.

Tveitdal, S. (2008). [Tallene må tolkes] Figures need to be interpreted. In Vetlesen, A. J. (ed.) [*Nytt Klima*] *A New Climate*. Oslo: Gyldendal.

Uri, H. (2006). [*De beste blant oss*] *The Best Among Us*. Oslo: Gyldendal, quote translated by Don Bartlett.
Urry, J. (2000). *Sociology Beyond Societies*. London: Routledge.
Urry, J. (2003). *Global Complexity*. Cambridge: Polity Press.
Urry, J. (2007). *Mobilities*. London: Polity.

Further reading

Abbott, C. (2008). *An Uncertain Future*. Report on Security and Climate Change for the Oxford Research Group.

Bhaskar, R. (1975/2008). *A Realist Theory of Science*. London and New York: Routledge.

Bhaskar, R. (1986/2009). *Scientific Realism and Human Emancipation*, especially Chapter 2. London and New York: Routledge.

Bhaskar, R. (1993/2008). *Dialectic: The Pulse of Freedom*. London and New York: Routledge.

Bhaskar, R. (1998). *The Possibility of Naturalism: A Philosophical Critique of the Contemporary Human Sciences*, especially pp. 173–174. Third Edition, London and New York: Routledge.

Bhaskar, R. (2002). *Reflections on meta-Reality*. New Delhi, Thousand Oaks, London: Sage.

Bhaskar, R. and Danermark, B. (2006). Metatheory, interdisciplinarity and disability research: a critical realist perspective. *Scandinavian Journal of Disability Research*, 8: 278–297.

Biggs, T. and Satterthwaite, R. (2005). *How to Make Poverty History: the Central Role of Local Organisations in Meeting the Millennium Development Goals*. London: IIED.

Commoner, B. (1971). *The Closing Circle: Nature, Man and Technology*. New York: Alfred Knopf.

Daly, H. E. (1993). Sustainable growth: An impossibility theorem. In H. E and K. N. Townsend (eds) *Valuing the Earth: Economics, Ecology, Ethics*, pp. 267–273, Cambridge, MA: Massachusetts Institute of Technology.

Dickens, P. (1996). *Reconstructing Nature: Alienation, Emancipation and the Division of Labour*. London: Routledge.

Flannery, T. (2007). *The Weather Makers: Our Changing Climate and What It Means for Life on Earth*. London: Penguin.

Gunderson, L.H. and Holling, C.S. (2001). *Panarchy: Understanding Transformations in Human and Natural Systems*. Washington DC: Island.

Henson, J. (2008). *The Rough Guide to Climate Change*, London: Penguin.

Hines, C. (2000). *Localisation: a Global Manifest*. London: Earthscan.

Høyer, K. G. and Næss, P. (2008). Interdisciplinarity, ecology and scientific theory: the case of sustainable urban development. *Journal of Critical Realism*, 7: 5–33.

IPCC (2007). *Climate Change 2007: Synthesis Report*. Contribution of Working Groups I, II and III to the Fourth Assessment Report of the Intergovernmental Panel on Climate Change [Core Writing Team, Pachauri, R.K and Reisinger, A. (eds)]. IPCC, Geneva, Switzerland.

Jackson, T, (2009). *Prosperity Without Growth*. London: Sustainable Development Commission.

Keen, M., Brown, V.A. and Dyball, R. (eds) *Social Learning in Environmental Management*. London: Earthscan.

Kriegler, E. *et al*. (2009). Imprecise probability assessment of tipping points in the climate system. In *Proceedings of the National Academy of Science in the United States*, February.

Millennium Ecosystem Assessment (2003). *Ecosystems and Human Well-Being: A Framework for Assessment*. Washington, DC: Island.

Næss, P. and Høyer, K. (2009). The Emperor's green clothes: growth, decoupling and capitalism. *Capitalism, Nature, Socialism*, 20 (3).

Nissani, M. (1997). Ten cheers for interdisciplinarity: the case for interdisciplinary knowledge and research. *Social Science Journal*, 34 (2): 201–216.

Odum, H.T. (2007). *Environment, Power and Society for the 21st Century*. New York: Colombia University Press.

Parker, J. and Wade, R. (eds) (2008) *Journeys Around Education for Sustainability*, London: South Bank University.

Sachs, W. (1999). *Planet Dialectics*. London: Zed Books.

UNDP (2007). *Fighting Climate Change: Human Solidarity in a Divided World*. Human Development Report 2007–2008. New York: Palgrave Macmillan. Full report and summary available for download from hdr.undp.org/en/reports/global/hdr2007-2008/

Wals, A.E.J. (ed.) (2007). *Social Learning: Towards a Sustainable World*. Netherlands: Wageningen.

Biographical notes on contributors

Carlo Aall
Carlo Aall received a graduate degree (CandAgric) in 1987 from the Norwegian University of Agriculture. He first worked as an environmental advisor for a small Norwegian municipality (1988–1990), but since 1990 has been employed as research associate and (since 2005) head of research at the Division for Sustainable Development at Western Norway Research Institute in Sogndal. He completed a PhD in Municipal Sustainable Development Policy at the University of Aalborg in 2000. He has published numerous reports – mostly in Norwegian – on sustainable development and climate policy, with a specific focus on the local level of government. He has also published and edited several scientific books in Norwegian on these issues – the most recent being a book summing up 10 years of experience in Norway on working with Local Agenda 21 which he co-edited with Professor William Lafferty at the University of Oslo. Furthermore, he has published 16 peer review articles in scientific journals and books, the most recent being an article on the scope of action for local climate policy published in Local Environment (2007).

Roy Bhaskar
Roy Bhaskar is perhaps best known as the originator of the philosophy of critical realism, and that later development of it which is the philosophy of meta-Reality. He is the author of many acclaimed and influential books and articles, including *A Realist Theory of Science* (1975), *The Possibility of Naturalism* (1979), *Scientific Realism and Human Emancipation* (1986), *Reclaiming Reality* (1989), *Philosophy and the Idea of Freedom* (1991), *Dialectic: The Pulse of Freedom* (1993), *Plato Etc.* (1994), *Reflections on meta-Reality* (2002) and *From Science to Emancipation* (2002). He is a co-editor of *Critical Realism: Essential Readings* (1998). He has lectured at universities and other institutions all around the world. He was the founding chair of the Centre for Critical Realism and is currently a World Scholar at the University of London Institute of Education.

Sarah Cornell
Sarah Cornell works on integrative socio-environmental research at the University of Bristol, where she directs a Masters programme in Earth System

Science and is currently engaged in multi-faculty research development on the human dimensions of global change. She is also the science programme manager for the UK NERC-funded programme QUEST (Quantifying and Understanding the Earth System, 2004–2010). Her background is in biogeochemistry, with a continuing interest in the perturbed global nitrogen cycle, and in environmental resource management. In recent years, she has become more engaged in use-orientated transdisciplinary research, with a particular focus on conceptualisations of humans in the Earth system.

Robert Costanza

Robert Costanza is the Gordon and Lulie Gund Professor of Ecological Economics and founding director of the Gund Institute for Ecological Economics at the University of Vermont. His transdisciplinary research integrates the study of humans and the rest of nature to address research, policy and management issues at multiple scales, from small watersheds to the global system. He is co-founder and past-president of the International Society for Ecological Economics, and was founding chief editor of the society's journal, *Ecological Economics*. He has published over 400 papers and 20 books. His awards include a Kellogg National Fellowship, the Society for Conservation Biology Distinguished Achievement Award, and a Pew Scholarship in Conservation and the Environment.

Cheryl Frank

Cheryl Frank was educated at the University of Illiinois, earning masters degrees in political science and journalism. She has completed extensive doctoral work in the fields of cultural studies and mass communications. A mother of two and grandmother of three children, in the early 1970s as a social activist and feminist organizer, she co-founderd one of the first women's domestic shelters in the USA. For many years she was a daily newspaper reporter for such publications as the Decatur [Illinois] *Herald & Review*, where she served as a bureau chief covering environmental and agricultural issues among others. She also worked as a legislative correspondent in the Illinois State Capitol for *The Chicago Daily Law Bulletin* and for daily newspapers. She was a writer for the *American Bar Association Journal* and has also done freelance work for many magazines. She has done extensive research on the position of women and Native Americans in the twentieth century. Her current interests include relating the philosophy of critical realism and meta-Reality to trends in British cultural studies and critical discourse analysis, especially in the fields of environmental education and peace studies.

Silvio Funtowicz

Silvio Funtowicz taught mathematics, logic and research methodology in Buenos Aires, Argentina. During the decade of 1980 he was a Research Fellow at the University of Leeds, England. He is now a member of the Institute for the Protection and Security of the Citizen (IPSC), European Commission – Joint Research Centre (EC-JRC). He is the author of *Uncertainty and Quality in Science for Policy* (1990, Kluwer, Dordrecht) in collaboration with Jerry Ravetz, and numerous papers in the field of environmental and technological risks and policy-related research. He has lectured extensively and he is a member of the editorial board of several publications and the scientific committee of many projects and international conferences.

John Hille

John Hille was educated in humanities, economics and earth sciences at the Universities of Trondheim and Oslo during two periods in the 1970s and 1980s, separated by 5 years of work as a farmhand. After working briefly for Statistics Norway in 1985–86 he joined the Project for an Alternative Future, an interdisciplinary research project whose aim was to explore alternative development paths for the Nordic countries in which social and environmental concerns were given priority. He was among the founders of the Ideas Bank, an offshoot of the former project whose central mission is to document and disseminate examples of best practice in sustainable development, and which has co-operated closely with Norwegian municipalities to help develop sustainable policies and practices. From 1991 until 2009 he divided his time between the Ideas Bank, teaching and research for the Centre for Development and the Environment at the University of Oslo and more recently the Western Norway Research Institute, and freelance writing. He is the author or co-author of over 40 reports on sustainable development issues, mainly in Norwegian. Having resigned his position at the Ideas Bank in 2009, he is now a full-time freelancer.

Karl Georg Høyer

Karl Georg Høyer is Professor and Research Director at Oslo University College, and is heading the college interfaculty and interdisciplinary research program *Technology, Design & Environment* (TDE). He holds an MSc in engineering sciences and a PhD in social sciences. The title of his PhD thesis was 'Sustainable Mobility – a Concept and its Implications', which he defended at Roskilde University in Denmark. For more than 30 years he has been a researcher and project leader for numerous research projects, including both Norwegian, Nordic and European projects. All this research has been interdisciplinary, mostly interconnecting technological, environmental and social science approaches and fields of knowledge. In these areas he has published several international scientific articles both in journals and in books, and has also contributed to and co-edited several scientific books in Norwegian. A later major Norwegian book is titled in English *Sustainable Development in Local Municipalities*, co-edited with Aall, C. and Lafferty, W. (Gyldendal Academic, 2002). Høyer was formerly Principal of Sogn og Fjordane University College, and Managing Director and Head of Research at theWestern Norway Research Institute. In 2008 he was one of the initiators behind the establishment of *Concerned Scientists*, Norway, where he also is a board member. This is an organization mobilizing some of the most highly profiled and renowned energy and climate change scientists in Norway in a common effort to radically change Norwegian climate change policies.

Hugh Lacey

Hugh Lacey is Scheuer Family Professor of Philosophy Emeritus at Swarthmore College, PA, USA, and Research Fellow in a project, 'The origins and meaning of technoscience', in the Philosophy Department, Universidade de São Paulo, Brazil. His recent publications include: *Is Science Value Free?* (1999), *Values and Objectivity in* Science (2005), and *Valores e Atividade Científica*, Volume 1 (2008) and Volume 2 (2009).

Maria Inês Lacey

Maria Inês Lacey received a PhD in Experimental Psychology from Universidade de São Paulo in 1973. She worked for several years with the American Friends Service Committee (an international NGO), utilizing popular education methods developed by Paulo Freire in workshops throughout the USA, and later as a teacher in an Adult Education Program in Philadelphia. She has also worked in translation, editing and writing.

Petter Næss

Petter Næss has been Professor in Urban Planning at Aalborg University, Denmark since 1998, with a part-time position at Oslo University College in Norway. Originally educated as an architect, Næss holds a doctoral degree (Dr. Ing.) in urban and regional planning. Næss has for many years carried out research into issues related to sustainable urban development, with a particular focus on the influence of spatial urban structures on travel behavior. Other main research topics are the philosophy of science and planning theory. In recent years his research has also addressed need analyses and assessment methods in planning and decision-making on large-scale transportation investments, driving forces of urban development, and ecological limits to economic growth within the housing and transportation sectors. His most recent books are *Urban Structure Matters: Residential Location, Car Dependence and Travel Behaviour* (Routledge, 2006), [*Bilringene og Cykelnavet*] *The Car Tyres and the Bike Hub* (Aalborg University Press, 2005, with Ole B. Jensen), and [*Fysisk Planlegging og Energibruk*] *Spatial Planning and Energy Use* (Tano/Aschehoug, Norway, 1997).

Jenneth Parker

Jenneth Parker has a background and first degree in philosophy and involvement in feminist and environmental movements for change. She has a Msc in social philosophy from the London School of Economics and an interdisciplinary PhD from Sussex University, linking ethics, critical realist philosophy of science and social movement theory in order to examine ecofeminist ethics. She is a former co-director of the Education for Sustainability distance learning masters programme at London South Bank University, developed by NGOs after the first Earth Summit of 1992. She has carried out consultancy in sustainability and education issues with a range of organisations including local government, NGOs such as WWF-UK, Science Shops Wales and internationally, with UNESCO, working on the Decade of Education for Sustainable Development. She is currently a research fellow at the Graduate School of Education, University of Bristol, working on interdisciplinarity and sustainability research, specifically on climate change. She has written on applied ethics and education for sustainability and has been writing on critical realism and sustainability since 1999.

Kjetil Rommetveit

Kjetil Rommetveit is a philosopher and a research fellow at the Centre for the Study of the Sciences and the Humanities, University of Bergen, Norway. His main research interests being ethics of science and technology and biopolitics, he currently works as researcher and project manager of the European FP7 project 'Technolife – a transdisciplinary approach to the emerging challenges of novel technologies: lifeworld and imaginaries in foresight and ethics'. Rommetveit holds a PhD from the same university, based on the dissertation 'Biotechnology: action and choice in second modernity'.

Roger Strand

Roger Strand is Professor and Director of the Centre for the Study of the Sciences and the Humanities, University of Bergen, Norway, and member of the National Committee of Research Ethics of Natural Science and Technology in Norway. He holds a PhD in biochemistry. His research mainly falls within the philosophy of natural science, environmental science and biomedicine, including research on the ethical and social aspects of bio- and nanotechnology. The focus of research is on the nature and significance of scientific uncertainty and complexity for environmental and health-related decision-making processes. A strong believer in interdisciplinary research and team work, most of his publications are co-authored. His more frequent co-authors are Sílvia Cañellas-Boltà, Dominique Chu, Ragnar Fjelland, Silvio Funtowicz, Jan Reinert Karlsen, Kamilla Kjølberg, Rune Nydal and Edvin Schei.

Index